I0540576

# Evolution's Stumbling Blocks

*Challenging Assumptions,*

*Exploring Purpose*

Chris Nitardy

© 2026 by Chris Nitardy

Published by **U R Loved Publishing LLC**

All rights reserved.

No part of this book may be reproduced, stored in a retrieval system, or transmitted in any form or by any means—electronic, mechanical, photocopying, recording, or otherwise—without prior written permission from the publisher, except for brief quotations in reviews or articles.

The author affirms that all content is original and does not infringe upon the legal rights of any person or work.

This publication is intended for general informational purposes only. The author and publisher make no warranties regarding the accuracy or completeness of the content and disclaim all liability for any loss, damage, or adverse consequences arising from its use. Nothing herein should be considered professional legal, financial, medical, or other advice. Readers should seek appropriate professional guidance as needed.

Due to the changing nature of the Internet, any websites or URLs mentioned in this book may have changed or may no longer be accessible.

The views expressed in this work are solely those of the author and do not necessarily reflect those of the publisher. The publisher disclaims responsibility for any opinion or interpretation expressed herein.

Unless otherwise indicated, Scripture quotations are taken from *The Holy Bible, New International Version*. Copyright © 1973, 1978, 1984 by International Bible Society. Used by permission. All rights reserved worldwide.

Paperback ISBN: 978-1-971002-69-9      U R Loved Publishing LLC
Ebook ISBN: 978-1-971002-67-5      U R Loved Publishing LLC
Hardback ISBN: 978-1-971002-68-2      U R Loved Publishing LLC

# Contents

# Acknowledgements

I would like to extend my sincere thanks to the following authors, publishers, and organizations for granting permission to reproduce excerpts from their works: the Institute for Creation Research, Answers in Genesis, Creation Ministries International, Walt Brown (*In the Beginning*), James Perloff (*Tornado in a Junkyard*), Bruce Malone (SearchForTheTruth.net), Vance Ferrell (*Science vs. Evolution*), Babu Ranganathan (freelance writer), John Sanford (*The Mystery of the Genome*), New Leaf Press, Master Books, Thomas Nelson Inc., Regnery Publishing, Prison Fellowship, Illustra Media, Grace to You, Ray Comfort, Zondervan, and all other referenced sources.

My deepest gratitude goes to Richard Gunther for graciously allowing me to feature his hilarious cartoon illustrations.

I am deeply grateful to my father, William, who planted the seeds that inspired this project, offered unwavering encouragement, and played a role in reviewing the manuscript. I am equally thankful to my stepmother, Diane, for her thoughtful contributions to the review process. I also extend heartfelt thanks to my sister, Ann, and her daughter, Amanda, for their valuable feedback during the review.

My deepest thanks to my dear wife, Katriina, and to all our wonderful children and stepchildren, who provided encouragement along the way.

I would also like to acknowledge the assistance of ChatGPT, a language model developed by OpenAI, which supported the revision process by helping refine the text.

Finally, I am deeply grateful to Jesus Christ for His abundant grace, forgiveness, and the ongoing personal relationship we share—made possible through His sacrifice. Thank You, Jesus.

# From Slimy Goo... Came Me and You

*In the beginning—there was nada,*
*No time, no light, no stellar cantata.*
*No stars to blaze, no planets to fling,*
*Not even a dino with half-formed wings.*

*Then boom! From zilch, a cosmic show,*
*The stars unfurled in swirling glow.*
*A galactic dance, life took its cue—*
*A riddle no mind could construe.*

*From Big Bang's dawn to monkeys with pens,*
*Rising, evolving as modern men.*
*From microbe to Mozart—atoms in strain,*
*A blind gospel enthroned in Darwin's domain.*

*Mindless forces brewed in nature's stew,*
*Yet somehow, thought emerged from the goo.*
*In primordial soup, they say we rose,*
*But how or why, nobody knows.*

No mind to steer, no compass, no plan—
From churning chaos, rose thinking man.

They say the proof is clear and grand—
But can it bridge life's jump to land?
From fish to frog—a baffling shift,
What force urged their limbs to lift?

What beckoned from water's deep,
To crawl ashore, then run and leap?

Great-grandpa's kin struck poses so spry,
Primates in tweed, swapping fur for a tie.
From lifeless stone to fearsome croc—
A bumbling, improbable stumbling block.

The tale they tell is sleek and grand,
But deeper truths slip through their hands.

Darwin's debut—a secular creed,
Scripture traded for worldly greed.
Some mock the cross, deny the soul,
Yet reason alone can't make us whole.

*"Intelligent Design?" they scoff, they flee—*
*But why recoil from divinity?*

*We gaze at stars, yet still we yearn—*
*For truths no science dares discern.*
*Did life arise from random chance,*
*Or was it sung in Love's first dance?*

*Entropy decrees, "All things decay!"*
*Yet Uncle Sam keeps Darwin in play.*
*The fossil record is patchy and thin—*
*Like Swiss cheese stretched on Darwin's grin.*

*Though hailed as truth, their logic snaps—*
*As Darwin's tale teeters on collapse.*

*Critics expose the cracks they find—*
*Thus Truth shines through the probing mind.*

*For not by chance, nor time, nor spin,*
*But by His Word did life begin.*

*Come, praise the God who formed it all,*
*Whose wisdom runs deeper than evolution's call.*
*His love restores what pride had erased—*
*And fills humble souls with truth and grace.*

*So when they claim we arose from the stew,*
*Smile and say:*
"Fearfully, wonderfully—*Jesus* made you."
*For by His breath the galaxies flew—*
*The Hand that shaped the stars formed you.*

# Monkey Business

# A Few Evolutionary Laughs

Why did dinosaurs leave so many fossilized footprints?

*They wanted to make a lasting impression.*

A little girl asked her father, "How did the human race appear?"

The father answered, "God made Adam and Eve, and they had children, and so all mankind was created."

Two days later, the girl asked her mother the same question.

The mother answered, "Many years ago, humans evolved from monkeys."

The confused girl returned to her father and said, "Dad, how is it possible that you told me the human race was created by God, and Mama said they evolved from monkeys?"

The father answered, "Well, dear, it's very simple. I told you about *my* side of the family, and your Mama told you about *hers*."

One day, the zookeeper noticed that the orangutan was reading two books—*The Bible* and *Darwin's The Origin of Species*.

Surprised, he asked the ape, "Why are you reading both those books?"

"Well," said the orangutan, "I just wanted to know if I was my brother's keeper or my keeper's brother."

---

Three freshman engineering students were sitting around talking between classes when one brought up the question of who designed the human body.

One student insisted that the human body must have been designed by an electrical engineer, because of the perfection of the nerves and synapses.

Another disagreed and exclaimed that it had to have been a mechanical engineer. "The system of levers and pulleys is ingenious!"

"No," the third student said, "you're both wrong. The human body was designed by an architect. Who else but an architect would have put a toxic waste line through a reproductive center?"

---

## Q: How do you tell the sex of a chromosome?

A: *Pull down its genes.*

# Introduction

# A Personal Search for Truth

*Between faith and science, a silent call,*

*To seek the truth beyond it all.*

## The Question That Changed Everything

My journey began many years ago when I first questioned the faith I had always taken for granted. I longed to see God—*truly* see Him—but His physical absence was deeply unsettling.

At the same time, the culture around me felt fractured and less innocent. I didn't stop believing, but faith was no longer effortless.

That shift extended to my experience of the church. The place I once turned to for hope now felt compromised.

I saw cultural decay creeping in, along with troubling apathy among believers and leaders alike. The growing gap between our former convictions and what we silently tolerated didn't just disturb me—it made me question whether any of it was real.

If the church—the place meant to embody truth—had lost its clarity, what could I trust?

Then the burning question hit me—one that kept me awake at night:

***Was I simply the product of random evolution, or intentionally created by a God who truly knows me?***

That question tugged at the loose threads of everything I believed.

It wasn't just about science or religion—it was about who I was, why I existed, and whether life had real purpose.

## A Moment That Changed Everything

One late afternoon in 2003, after a Promise Keepers event, the tension inside me finally broke.

I stepped onto my deck—desperate—and asked God, out loud, to show me He was real. In that moment, I even asked for a sign in the sky.

Then I looked up and saw what looked unmistakably like the word *"God"* formed in the clouds. It lingered for about ten seconds. Then, slowly, it faded.

I've only shared this with a few people before, but now I believe it's time to speak of it more openly.

I stood there—stunned—as tears rolled down my face.

Was it coincidence? Imagination? Or something more?

That moment shook me, but it didn't erase all my doubts—I'm naturally skeptical.

Even after what I saw, I knew emotion alone wouldn't be enough. I needed something firm, so I turned to Scripture and began examining the evidence for myself.

## A Search for Something Solid

I wasn't seeking debate—I wanted to know what was real. Diving into Darwin's theory, the Bible, and competing worldviews took me from crisis to spiritual awakening.

I had to rethink everything—origins, morality, religion, and God Himself.

It didn't give me all the answers, but it made me ask better questions.

This book explores the stumbling blocks I encountered and the moments when evidence, faith, and human assumptions collided.

As you walk this path with me, you may find that those same moments of tension can become opportunities—for clarity, for insight, and even for transformation.

God had always been part of my life, but through this process, He became profoundly real and personal.

Before that shift, my understanding of Scripture was limited. So I made a decision: I would read the Bible cover to cover—no commentaries, no outside filters—just me and the text.

At the same time, I approached evolutionary theory with fresh eyes, willing to follow the facts wherever they led.

Along the way, I discovered that true understanding often comes not from quick answers, but from the willingness to sit with uncertainty and explore it deeply.

What you're about to read is the fruit of that search.

Without formal academic titles, I was free to question both science and Scripture without the pressure of institutions or traditions.

This work isn't anti-science. It's a call for openness, honest exploration, and the courage to question even widely accepted ideas.

Whatever your background—skeptic, believer, or seeker—you're invited to examine the evidence.

If you're confident in evolution, these pages are still for you—not to argue, but to explore.

*What if the complexity of life, the assumptions of science, and the deeper questions of existence all point beyond evolution—to something more?*

## Why This Matters

In a distracted world, it's easy to overlook life's deeper questions. But ignoring them doesn't make them disappear—it only leaves us shaped by unexamined assumptions.

What we believe about our origins is more than a scientific position; it's the lens through which we see everything else—morality, purpose, even identity.

If life is merely the result of random processes, then morality and meaning rest on unstable ground. Right and wrong become subjective, and valuing others can become optional—especially when inconvenient.

But if we were intentionally created, every life carries inherent worth. Our choices, relationships, and even our suffering gain significance.

This isn't about winning an argument—it's about seeking understanding. With curiosity, humility, and integrity, we'll pursue one essential question:

*What does the evidence really show?*

## Evolution: Science, Philosophy—or Both?

Is evolution purely scientific?

Public education often presents it—including the origins of life and the cosmos—as the only valid narrative, dismissing other perspectives. This raises a fair question: is this objective science, or are philosophical assumptions also at play?

Modern evolutionary theory encompasses not only how species change but how life itself began. The debate isn't merely biological; it's about whether life arose by chance or by design.

This reveals a deeper divide: the Bible's *"In the beginning, God created..."* versus the secular view of spontaneous emergence from nothing.

Some argue that promoting only one narrative in schools raises both ethical and constitutional concerns. We'll examine this further in *Stumbling Block #15: Evolution in Schools—Is There a Bias?*

As the saying goes, *"Ideas have consequences."* Darwinian theory has become so dominant in academia that it often discourages honest debate. Yet the evidence remains—waiting for fresh examination.

It's never too late to ask hard questions.

The following illustration explores how evolution, often presented as settled science, may carry philosophical assumptions beneath its scientific surface.

EVOLUTION IS PHILOSOPHY
PRETENDING TO BE SCIENCE

This book delves deeper into evolution's core claims. What once seemed unquestionable may now be open to fresh scrutiny. As we explore the evidence, your views on creation, evolution, religion, and even the Bible may begin to shift.

## Science and Faith—Conflicting Worldviews

The tension between science and faith often stems from their foundational assumptions. Though frequently portrayed as opposites, the relationship is more nuanced.

At the core:

- **Evolution** starts with a naturalistic assumption—that life must be explained without invoking the divine.

- **Creation** begins with a theistic assumption—that life is the result of intentional design by a Creator.

Both approaches draw from scientific evidence but interpret it through different lenses. This challenges the idea that evolution is purely objective or free from belief-based assumptions.

Consider this paraphrased summary from George Wald's 1954 article *The Origin of Life* in *Scientific American*:

> *When it comes to the Origin of Life, there are only two possibilities: creation or spontaneous generation. There is no third way. Spontaneous generation was disproved one hundred years ago, but that leads us to only one other conclusion, that of supernatural creation. We*

*cannot accept that on philosophical grounds; therefore, we choose to believe the impossible: that life arose spontaneously by chance![1]*

This highlights a key point: our worldviews often shape our conclusions—even more than the evidence itself.

## When Ideology Wears a Lab Coat

Philosophical bias can fuel intellectual censorship, turning Darwinian evolution into an untouchable pillar within many scientific circles. The documentary *Expelled: No Intelligence Allowed* reveals how dissenting voices are often marginalized.[2]

As Dr. Rebecca Keller, a biophysical chemist, states:

*If we, as scientists, are not allowed to question, ponder, explore, and critically evaluate all areas of science, then we are operating in a mode completely antithetical to the very nature of science.[3]*

True science demands freedom—the freedom to ask tough questions, challenge assumptions, and weigh competing ideas without ideological pressure. All perspectives—including creation and evolution—deserve that freedom.

## A Rising Tide of Doubt

If Darwinian evolution is truly settled, why are so many scientists questioning it? Despite popular belief, a growing number of researchers express skepticism. More than a thousand PhD-level scientists have signed the ***Dissent from Darwin* petition**, stating:

> *We are skeptical of claims for the ability of random mutation and natural selection to account for the complexity of life. Careful examination of the evidence for Darwinian theory should be encouraged.*[4]

Far from fringe, this movement includes scientists from leading universities and diverse fields. They're not rejecting science—they're calling for honest evaluation.

Consider these voices:

**Dr. Stanley Salthe**, a zoologist who once specialized in evolutionary theory, wrote:

> *Darwinian evolutionary theory was my field of specialization in biology. I wrote a textbook on the subject thirty years ago. Meanwhile, however, I have become an apostate from Darwinian theory and have described it as part of modernism's origination myth.*[5]

**Douglas Axe**, a molecular biologist and Biola University professor, said:

> *Because no scientist can show how Darwin's mechanism can produce the complexities of life, every scientist should be skeptical. The fact that most won't admit to this exposes the unhealthy effect of peer pressure on scientific discourse.*[6]

These aren't isolated opinions. They reveal a growing undercurrent of doubt—challenging the idea that Darwinian evolution is beyond question. Too often, this conversation is ignored or dismissed in mainstream education and media.

What's really at stake is not just the evidence—but how we interpret it.

**Ken Ham**, founder of *Answers in Genesis*, sums it up:

> *The controversy is not religion versus science, but the science of one religion against the science of another religion.*[7]

His point isn't to dismiss science, but to show that everyone begins with foundational assumptions that shape what they accept as reasonable or true. The central question becomes:

**Which worldview best explains the data and offers the most coherent understanding of life's origins?**

This debate is just beginning—and the full story is waiting to be uncovered.

## Erosion of Truth

Darwinian evolution's reach extends beyond biology—it reshapes how society views truth, morality, and meaning. If life arose through random, unguided processes, objective facts become elusive, replaced by personal preference.

Without a transcendent foundation, values such as justice, dignity, and notions of right and wrong erode, yielding to convenience and ideology. We see this in today's fractured culture—where misinformation spreads, trust wanes, and core principles are discarded. Like Pilate, the world asks, "What is truth?"—yet often does so without seeking an answer.

Many trace this drift toward relativism to a purposeless worldview rooted in Darwinian theory. *When objective reality fades, unity and shared purpose fade with it.*

## Who Are We Really?

How we view our origins profoundly shapes our identity and sense of purpose. If we see ourselves as the product of random chance—assembled without design—then meaning, morality, and hope become self-created. Life may feel purposeless, and death final.

If we believe we were intentionally created by God, life carries inherent value and meaning. We are known, loved, and designed for a purpose. Even suffering and death take on new significance through the lens of eternity.

These beliefs shape how we answer life's deepest questions:

- **Who am I?**
- **Why am I here?**
- **What is the purpose of my life?**

- **Where am I going when I die?**

Consider two contrasting perspectives:

An atheist might say:

> *I'm a collection of meaningless molecules assembled by chance from a cosmic explosion that took place billions of years ago. There is no ultimate purpose to my life, and when I die, all the atoms in my body will simply be recycled into the vast, dark, cold universe.*

A Christian might say:

> *My life is full of meaning and purpose, for I am a child of the living God, and the Holy Spirit resides within me. Every amazing cell in my body was carefully crafted and woven together by the omnipotent Creator of the universe. Death is not the end, but simply a new beginning, for Jesus conquered sin and death. My soul longs to worship and commune with God for the rest of eternity.*

Not all atheists see life as meaningless, and not all believers are always confident in their faith. With that in mind, ask yourself:

- Which perspective resonates more with you?
- What is the foundation of your worldview—and how confident are you in it?

It's worth noting that not all who accept evolution reject God; some hold to *theistic evolution*—the idea that God used evolutionary processes in His creative work. We'll explore this further in Stumbling Block #16: *The Bible vs. Evolution*—does it truly harmonize science and Scripture, or compromise both?

Let's compare these basic assumptions:

**Naturalistic Evolution Teaches:**

- Life came from non-life (a process never observed).
- Order and information arose from chaos.
- Consciousness emerged from unconscious matter.
- Morality developed in a universe without meaning.
- Everything came from… nothing.

**The Bible Teaches:**

- Life only comes from life (consistent with science).
- Order and design reflect a Designer.
- Consciousness comes from a conscious Creator.
- Morality is rooted in God's unchanging character.
- Everything was made with purpose—by Someone.

Only one view aligns with what we observe: order, design, and moral intuition. Can a blind, purposeless process explain this, or does evidence point to intentional design?

This isn't just about how life began—it's about what life means.

Wherever you stand—skeptical, searching, or confident—examine the evidence with fresh eyes.

Let's keep asking questions.

Let's keep thinking critically.

And above all, pursue the *Truth*—in love, because *Truth* is worth everything.

## Cracks in the Foundation

Since *The Origin of Species* was published in 1859, Darwin's theory has been taught as the cornerstone of modern biology—an unshakable pillar of science. For many, it seems beyond question: settled, proven, essential.

It's a familiar story, heard since childhood, rarely questioned. But the reality is more complex.

Despite wide acceptance, Darwin's theory has significant gaps. Questions remain about the fossil record, the origin of life, and the limits of mutation and natural selection. The deeper you look, the more uncertainty emerges.

In some circles, evolution functions less like science and more like cultural orthodoxy—embraced not just for evidence but because alternative ideas are off-limits.

Questioning it can cost credibility or even a career.

True science, however, welcomes honest inquiry.

## A Theory Under Pressure

Darwin's theory rests on underlying assumptions and continues to face many unresolved problems, sparking debate among scientists, philosophers, and theologians alike. The following chapters explore some of the most widely discussed and controversial challenges:

- Gaps in the Fossil Record
- The Cambrian Explosion
- Dinosaurs
- Limits of Mutation and Natural Selection
- Irreducible Complexity
- Origin of Life
- Evidence for a Global Flood
- A Fine-Tuned Universe
- The Bible vs. Evolution

## Beyond the Evidence

Science doesn't operate in a vacuum. What we believe about origins shapes how we see truth, morality, and the meaning of life. Each chapter will challenge assumptions and offer fresh perspectives—grounded in both scientific inquiry and a biblical worldview.

Wherever you stand—skeptic, seeker, believer—you're welcome in this conversation. The focus isn't to shame or argue, but to spark deep thinking, honest questions, and an open pursuit of evidence. The goal is understanding—pursued with humility and love.

Love doesn't mean accepting what's harmful, but it does call us to treat others with dignity and respect. True love speaks honestly, even when it's hard. Throughout these pages, you'll find moments of humor, sarcasm, and candid critique—meant to illuminate, not to offend.

You don't need a PhD—just curiosity and courage to ask:

## What if we've been missing something?

Because what you believe about your origins shapes everything about your destiny.

Looking back, I see something powerful:

> *What was intended for harm, God used for good. —*
> *Genesis 50:20*

The struggle wasn't easy, but it opened the door to a deeper, personal understanding of God and His purposes.

Let's move forward—with honesty, courage, and an unyielding pursuit of truth.

# Micro vs. Macro Evolution

# What's the Difference?

*A little tweak, a color change*

*And suddenly, it's all so strange!*

*Can tiny shifts create something new,*

*Or are these changes just small and few?*

What does real science reveal? Evolution is a cornerstone of modern biology, explaining how species adapt and diversify. Yet a central question remains: can small genetic changes truly accumulate to produce entirely new life forms over time?

## Defining Microevolution and Macroevolution

**Microevolution** refers to small, observable changes within a species. These adaptations—variations in color, size, or shape—help improve survival but stay within existing genetic boundaries.

**Macroevolution**, by contrast, claims that many such changes accumulated over time can eventually lead to the rise of entirely new species.

While microevolution is well-documented, macroevolution remains a theoretical framework, with its conclusions based largely on inference, fossil interpretation, and computer models—rather than direct, repeatable observation.

## Observable Examples of Microevolution

Pesticide resistance in insects is a textbook case. When exposed to pesticides, most insects die, except for a few carrying genetic variations that offer resistance. These survivors reproduce, passing on the resistant trait. Over time, resistant insects dominate. This is adaptation, not transformation; they remain the same kind of insect.

Darwin's **Galápagos finches** show a similar pattern. Depending on their environment, these birds display variations in beak shape and size to access available food. While these changes offer survival advantages, the birds remain finches—not hawks, hummingbirds, or any other kind.

### THE FINCH AS IT HAS EVOLVED OVER 6000 YEARS

| 6000 years ago | 5000 years ago | 4000 years ago | 3000 years ago | 2000 years ago | 1000 years ago |

AS YOU CAN SEE, THERE IS TREMENDOUS SUPPORT FOR THE THEORY OF EVOLUTION!

These examples highlight a crucial point: microevolution enables adaptation within a species, but it fails to demonstrate the kind of large-scale transformation macroevolution requires.

## The 98% Similarity Debate—and the Daffodil

It's often said that humans and chimpanzees share 98% of their DNA, suggesting close evolutionary ties. But does that tell the whole story? Factoring in gene structure, regulation, and expression, the actual similarity drops closer to **95%**.[1]

Even small genetic variations hold profound significance. As evolutionary biologist Richard Dawkins states:

> *The one percent difference, in absolute terms, is still a mighty quantity of DNA. It is enough to account for all the differences between humans and chimpanzees, however large and unbridgeable they may seem.*[2]

This means a seemingly small percentage can encompass deep biological distinctions. The real question isn't how many letters we share, but what those letters do. Tiny genetic differences can lead to enormous changes. While 95–98% sounds impressive, the impact of those differences is massive.

And if we're taking the 98% claim seriously, consider this: humans share about 35% of their DNA with daffodils. So—are we part flower now?

3

# Humans and daffodils possess a 35% genetic similarity.

This humorous thought makes a serious point: genetic similarity doesn't equal common ancestry. It often reflects shared cellular functions common to all living organisms. Just as two houses might use similar bricks, plumbing and wiring without being related, genetic similarity may point to common design—not descent.

## The Genetic Wall Between Species

Genetic research reveals that microevolutionary changes don't bridge the vast genetic gaps between species. Genes influence traits like hair or coat color but don't contain instructions to transform a dog into a cat. The fundamental genetic differences between species are substantial, and no observable pathway exists for microevolution to cross these boundaries.

As Vance Ferrell puts it:

> *The genetic makeup within the chromosomes forms a barrier, a literal wall of separation between one species and another.*[3]

This genetic wall suggests microevolution cannot account for the origin of new species—leaving macroevolution without a solid empirical foundation.

## Insights from Experts

**Dr. Lane P. Lester**, Professor of Biology and co-author of *The Natural Limits to Biological Change*, has studied the limits of genetic variation for decades. In his article *"Genetics: No Friend of Evolution,"* he writes:

*Genetics and evolution have been enemies from the beginning.[4]*

Lester argues that genetic variation operates within predetermined limits, encoded in DNA, preventing the large-scale transitions macroevolution requires.

Dr. John Sanford, a former Cornell geneticist and author of *Genetic Entropy*, echoes this:

*Macroevolution is an unsubstantiated assumption. There is no evidence that it has ever occurred, and a theory based on an assumption is no better than the assumption itself.[5]*

Sanford points out that mutations tend to degrade genetic information, not create new complexity. Similarly, Dr. Georgia Purdom, a molecular geneticist, states:

*While **microevolution** (small changes within a species) is clearly observable, there is no compelling evidence that such changes can lead to the origin of entirely new species or higher taxa. The genetic evidence does not support the large-scale evolutionary changes required for **macroevolution**.[6]*

Their work, along with others, supports the idea that variation occurs within created limits—a view closely aligning with the biblical account.

## Microevolution and the Bible's "Kinds"

The Bible teaches that creatures were made according to their "kinds," allowing for variation within those kinds but offering no evidence that one kind transforms into another.

This also sheds light on human diversity. If all humans descended from Adam and Eve, where did different races come from? Writer Babu G. Ranganathan explains:

> *Young people, and even adults, often wonder how all the varieties or 'races' of people could come from the same human ancestors. Well, in principle, that's no different than asking how children with different color hair (i.e., blond, brunette, brown, red) can come from the same parents who both have black hair.*
>
> *Just as some individuals today carry genes to produce descendants with different hair and eye colors, humanity's first parents possessed the genetic potential to produce all the varieties and races of people. While we may not carry the genes to produce every human variety today, humanity's first ancestors did possess such genes.*
>
> *All human varieties carry genes for the same basic traits, but not all humans carry every possible variation of those genes. For example, one person may carry*

*several variations of the gene for eye color (i.e., brown, green, blue), while another may possess only one variation (i.e., brown). Thus, both will have different abilities to influence the eye color of their offspring.*

Ranganathan goes on to explain:

*Some parents with black hair, for instance, may produce children with blond hair, but their blond children (who inherit only recessive genes) will not have the ability to produce children with black hair unless they mate with someone else who has black hair. If the blond descendants only mate with other blondes, then the entire lineage will remain blond, even though the original ancestor had black hair.*

*Unless Nature possesses the intelligence and ability to perform genetic engineering (to construct entirely new genes rather than merely producing variations and new combinations of existing genes), macroevolution will never be possible in nature.*

*We have varieties of dogs today that did not exist a couple of hundred years ago. The genes for these varieties have always existed within the dog species, but they simply lacked the opportunity for expression until the right conditions arose. The genes themselves*

*did not evolve! What we commonly refer to as
"evolution" is, in reality, the physical expression of
already existing genes. This exemplifies microevolution
(horizontal evolution) in nature.[7]*

The genetic boundaries observed in microevolution mirror the
Biblical account, which teaches that God created organisms
"according to their kinds." From this perspective, Adam and Eve
were designed with a rich genetic infrastructure, a gene pool
containing all the genetic information needed to produce the full
range of human diversity. Likewise, other created "kinds" were
genetically equipped for variation within their category. This
reinforces the idea that variation arises from built-in genetic
potential, not from the development of entirely new genetic
material.

The Bible repeatedly emphasizes this concept:

*God made the wild animals according to their kinds,
the livestock according to their kinds, and all the
creatures that move along the ground according to
their kinds. And God saw that it was good. — Genesis
1:25*

## The Micro vs. Macro Confusion

Scripture never teaches that life evolved from a single common ancestor. Instead, it presents distinct lineages—consistent with microevolution and limited variation within created kinds.

Much confusion arises from blurring the line between micro and macroevolution. Small, observable changes are often used to support massive, unobservable transformations—a false extrapolation beyond what the data shows.

Microevolution is backed by observable science, while macroevolution relies on interpretation rather than direct evidence.

This overarching distinction is critical to understanding the rest of our discussion. As this book unfolds, the truth of this will become undeniably clear.

A fair look at the evidence may surprise readers—including some evolutionists—by revealing that macroevolution falls short of the standards of rigorous science and the scientific method.

A former atheist illustrates how **science**—not religion—can lead people away from evolution and toward belief in a divine creator:

> *My road to atheism was paved by science...but, ironically, so was my latter journey to God. — Lee Strobel*[8]

## Conclusion: The Limits of Genetic Change

*Microevolution*—small, observable changes within a species—is a well-documented and universally accepted fact of biology. Macroevolution, by contrast—the idea that countless small changes can accumulate over time to create entirely new kinds of creatures—remains a theoretical model, not an observed phenomenon.

While evolution technically addresses changes within existing life, many modern discussions blur it with *abiogenesis*—the idea that life sprung from non-living matter. For clarity, both topics will be addressed, as they're often intertwined in science and education.

Recognizing the limits of microevolution raises a deeper question:

**Does life's complexity point to random chance—or intentional design?**

And if design… what kind of Designer?

That question is where our journey begins

Next, we'll explore what the fossil record really says—*and what it doesn't.*

.

# Fossilized Silence

# What the Fossil Record Isn't Saying

*If a story is told of a frog turning into a prince overnight,*

*We'd call it a fairy tale.*

*But when the story is stretched over millions of years,*

*Textbooks call it science.*

## Evolution's Greatest Evidence—or Its Greatest Embarrassment?

We're often told the fossil record is evolution's greatest evidence—a slow, steady chronicle of life transforming from simple to complex. Textbooks present it as a settled story: from fish to amphibians, reptiles to birds, mammals to man.

But when we examine the rocks, the story changes.

Instead of gradual transitions, we find sudden appearances—fully formed creatures with no traceable ancestors.

Rather than a clear and continuous record, the fossil archive looks more like Swiss cheese, riddled with holes where transitional forms should be.

If the previous chapter revealed the *limits* of evolution, this one confronts the *evidence* head-on. What we find in the ground doesn't confirm Darwin's theory—it challenges it.

And after more than a century of digging, the silence is deafening.

## Darwin's Dilemma

This fossil silence isn't just a modern observation—it's a problem Darwin himself anticipated.

In *The Origin of Species*, he candidly wrote:

> *Why then is not every geological formation and every stratum full of such intermediate links? Geology assuredly does not reveal any such finely graduated organic chain; and this, perhaps, is the most obvious and serious objection which can be urged against my theory.*[1]

Darwin hoped the gaps were temporary—that future fossil discoveries would eventually fill in the story. Yet more than 150 years later, and with billions of fossils unearthed, the silence has only grown louder.

Author Luther Sunderland captured the magnitude of this issue when he wrote:

*Now, after over 120 years of the most extensive and painstaking geological exploration of every continent and ocean bottom, the picture is infinitely more vivid and complete than it was in 1859. Formations have been discovered containing hundreds of billions of fossils, and our museums are filled with over 100 million fossils of 250,000 different species. The availability of this profusion of hard scientific data should permit objective investigators to determine if Darwin was on the right track. What is the picture which the fossils have given us? The gaps between major groups of organisms have been growing even wider and more undeniable.[2]*

"I'm sorry sir- they haven't found any yet."

The fossil record does not support a slow, step-by-step transformation of life. It shows sharp breaks, sudden appearances, and a conspicuous lack of intermediate forms.

Instead of a continuous evolutionary timeline, what we find is a film reel with the middle frames cut out—just abrupt beginnings and sudden endings, but no story in between.

Some have tried to explain these fossil gaps with a theory called *punctuated equilibrium*—suggesting that evolution happens in rapid bursts, leaving few fossils behind.

But if that idea solves anything, it raises even more questions—especially when we look at the most dramatic fossil gap of all: the sudden appearance of complex life in the Cambrian layer.

We'll tackle that next in Stumbling Block #3: *The Cambrian Explosion—Evolution's Big Bang.*

## Lost in Transition: Evolution's Identity Crisis

Evolution makes some bold claims: that birds evolved from dinosaurs, whales from hoofed mammals like cows, bats from tiny shrew-like creatures, and fish slowly moved onto land.

The claim is that countless small changes, spread over vast stretches of time, gradually remodeled one creature into something entirely new. But these aren't just minor tweaks—they're massive, coordinated transformations. They are complete overhauls of bones, organs, muscles, and entire body systems.

Yet, when we look at the fossil record—the geologic archive of life's history—we don't see the gradual remodeling in progress. We find fossils of what came before—and what came after. But the crucial in-between stages are mysteriously absent.

## Zooming in on Evolution's Biggest Mysteries

Let's dive into some of the rock layers' most baffling puzzles—starting with one of evolution's biggest head-scratchers: the sudden appearance of bats.

## Bat Out of Nowhere

Evolution claims bats sprang from small, ground-dwelling mammals—something like shrews. But bats are no ordinary mammals. They wield wings that rival any aircraft design and use a sophisticated sonar system to hunt insects midair in total darkness.

If this transformation were truly gradual, where are the fossils of half-winged, proto-flying mammals? In reality, bats emerge fully formed, masters of the night skies—as if evolution skipped flight school and went straight to graduation.

That's a... **bat-tacular** leap, indeed.

And let's be honest—if bats really developed sonar by trial and error, imagine how many cave walls they had to slam into before getting it right.

That's not natural selection—it's natural demolition.

Yet sonar requires a fully integrated auditory, neurological, and anatomical system. There's no evidence for "half-sonar." It's either fully functional—or completely useless.

If sonar stumps evolution, just wait until we get to the bugs.

## Bugged by Evolution

Insects—millions of species buzzing, crawling, flying—should flood the fossil record with transitional forms. Yet ancient insects appear suddenly, fully formed and unchanged for "millions of years," like a copy-paste in nature's history book.[3]

Take *Goniurellia tridens*—a fruit fly with an astonishingly detailed image of an ant on each wing. Not a vague shape, but an actual, anatomically precise ant, complete with legs, antennae, and a segmented body. It's so precise, some mistake it for AI-generated art. But it's real.

Try explaining that through blind mutation. This isn't nature stumbling—it's design showing off.

Such intricate design points to a creative hand behind the curtain.

And insects aren't alone in their sudden arrival. When we turn from the skies to the seas, the fossil record gets even murkier. *Just ask the fish—if they could talk.*

## Something's Fishy in the Fossil Record

Fish turning into land animals is often hailed as a textbook evolutionary triumph. Tiktaalik, the supposed "missing link," is frequently cited—but it shows up with limb-like fins and a flattened head, with no clear evolutionary trail before or after.

But moving from water to land isn't a simple tweak—it requires a coordinated overhaul of bones, muscles, lungs, and internal systems. And yet, the fossil record doesn't show a sequence—just snapshots.

The *coelacanth* was once celebrated as another key evolutionary bridge. Though extinct for 70 million years, it was believed to have lungs, a large brain, and be halfway to walking. That is, until 1938—when it was discovered alive off the coast of Madagascar.

Turns out, the coelacanth wasn't climbing onto land—it was just swimming. Fully aquatic. No lungs. No walking limbs. No landward ambitions. Dozens more have been found since— unchanged and uninterested in evolving.

So what happened? The story evolved faster than the evidence. A so-called "transitional form" became a living fossil—undermining the narrative it was supposed to support.

Why were such claims taken so seriously in the first place? Because for many, the issue isn't just fossils—it's a worldview.

Evolutionary science often begins with a philosophical commitment: life must be explained without invoking design. From that lens, even fragmentary or contradictory evidence gets bent to fit the story. Gaps become mysteries. Sudden appearances are labeled "rapid evolution." Problematic fossils are recast to preserve the narrative.

In the end, it's not just about what the rocks show—but what some believe they must show. And that shapes how the fossil record is read, taught, and defended.

Enter the whale.

## From Moo to Deep Blue: A Whale of a Tale

Evolutionists propose whales evolved from land mammals—perhaps from cows. But making that leap would require a complete anatomical overhaul:

- Front legs becoming paddle-shaped flippers
- Hind limbs shrinking or vanishing without disrupting internal organs

- Ears rewiring for underwater hearing
- Lungs adapting to deep-sea pressure and diving
- And most bizarre—nostrils migrating up the skull to form a blowhole

That's not evolution by baby steps—it's a full-body engineering project. One wrong mutation, and the creature's stuck in evolutionary limbo: no longer fit for land, not yet built for sea.

Transforming a car into a submarine wouldn't happen by accident—nor would turning a land mammal into a whale. Every system—from breathing and hearing to locomotion and reproduction—would need to be redesigned in sync, or the transition fails.

Fossils like *Pakicetus* are often paraded as transitional forms, but debates continue about their exact evolutionary placement. In reality, *Pakicetus* had legs, lived on land, and looked more like a dog than a dolphin. The evolutionary narrative tries to stretch it into a whale ancestor—but the fossil tells a different story.

Dr. Jonathan Sarfati points out: if the pelvis shrinks too early, walking fails. Too late, and tail-powered swimming crushes internal organs. That's not just unlikely—it's biologically unworkable.[4]

The layers of sedimentary rock don't show whales gradually swimming into existence. They appear suddenly—fully aquatic, fully functional, and ocean-ready.

Their massive size and complex features defy evolutionary expectations and cast serious doubt on any gradual land-to-sea transition.

From the ocean depths to the skies above, the story of bird evolution presents another fascinating puzzle.

## Feathers Without a Flight Plan

Archaeopteryx has long been hailed as the iconic "missing link" bridging the gap between dinosaurs and birds. Yet, despite thirteen remarkably preserved fossils, it doesn't look like a halfway creature struggling to take off.

Instead, Archaeopteryx appears as a fully feathered, flight-ready bird—complete with wings designed for agile flight and feet built for perching. Yes, it had claws and teeth, but some modern birds sport those too.

Renowned ornithologist Dr. Alan Feduccia, a non-creationist and skeptic of the dinosaur-to-bird theory, cuts straight to the chase:

> *Paleontologists have tried to turn Archaeopteryx into*
> *an earth-bound, feathered dinosaur. But it's not. It is a*

*bird, a perching bird. And no amount of 'paleobabble'*
*is going to change that.*[5]

Evolutionists argue that feathers first evolved for display or insulation before being co-opted for flight—a sort of evolutionary repurposing.

But think about it: that's like inventing jet engines before figuring out how to build an airplane. Avian flight demands an intricate design involving hollow bones, advanced respiration, a keeled sternum for muscle attachment, and precise aerodynamic control. These features don't emerge from trial and error—they require coordination on a scale that defies chance.

Yet, the fossil record refuses to tell this gradual story. Rather, it presents a stark divide: dinosaurs roaming on the ground, and birds already airborne.

These feathered creatures didn't taxi down an evolutionary runway—they launched straight into the skies.

*Birds of a feather may flock together—*

*but the idea that they hatched from dinosaurs?*

*That theory still doesn't fly.*

## What About the Missing Links?

A common rebuttal to intelligent design is that the fossil record is incomplete—that "missing links" aren't missing forever, just waiting to be found. It's a popular fallback: *"Just give it time."* But the very idea of a "missing link" is fundamentally misleading.

If evolution were a slow, continuous process, we'd expect at least a modest abundance of transitional fossils—not just a few rare examples. Critics often say, if evolution were true, we should be *"tripping over"* them.

Instead, we find the opposite: fully formed creatures appearing abruptly and remaining unchanged.

The gaps aren't occasional—they're everywhere. Not just between major groups, but across the board. After over a century of excavation, the geologic record reveals abrupt appearances and evolutionary dead ends—not the steady transitions predicted.

At some point, the absence of evidence is no longer a delay, but a verdict.

The silence of the fossils doesn't whisper of future finds—it shouts of a broken narrative.

## Dr. Colin Patterson's Shocking Admission

Here's a real eye-opener for you:

Dr. Colin Patterson, a former senior paleontologist at the British Museum of Natural History and one of the world's foremost authorities on the fossil record, made a jaw-dropping admission. In a letter responding to a creationist who questioned the absence of transitional fossils in his book, Patterson replied:

> *I fully agree with your comments about the lack of direct illustration of evolutionary transitions in my book. If I knew of any, fossil or living, I would certainly have included them... I will lay it on the line—there is*

*not one such fossil for which one could make a watertight argument.*[6]

That's a bombshell in the world of paleontology—an unexpected *mic-drop moment.*

After decades of research, one of the top voices in paleontology admitted: there's no solid fossil evidence for clear evolutionary transitions.

If these "missing links" are central to the story, the question remains—*why are they still missing?*

## The Gaps in the Record: Stones That Cry Out

The absence of transitional fossils isn't a minor oversight—it's a gaping hole at the heart of evolutionary theory. If life evolved through countless small steps over millions of years, the rock layers should be overflowing with creatures caught mid-transformation.

Even some evolutionary scientists have candidly acknowledged the issue. Zoologist David B. Kitts remarked:

*Despite the bright promise that paleontology provides a means of 'seeing' evolution, it has presented some nasty difficulties for evolutionists, the most notorious of which is the presence of 'gaps' in the fossil record.*

*Evolution requires intermediate forms between species, and paleontology does not provide them.*[7]

These aren't isolated anomalies. Even after uncovering hundreds of millions of fossils, the expected transitions are still conspicuously absent.

What we find in the rocks are fully developed, functional life forms—creatures that arrive in the record suddenly, without precursors, leaving no evolutionary breadcrumb trail behind.

And that brings us to a striking echo from Scripture. In Luke 19:40, Jesus says:

**I tell you, if they keep quiet, the stones will cry out.**

It's a profound image. If truth is silenced, *creation itself will speak.* Perhaps these fossil gaps aren't accidents of nature, but intentional signposts—not reminders of missing data, but of a missing paradigm.

Maybe the missing pieces… are the message. *The earth isn't silent—it's testifying.*

## View from a distance and see what it says.

Imagine it: the stones themselves crying out—not with words, but by what they refuse to reveal. They don't show life evolving through slow, accidental steps. What they hold—and what they don't—speaks volumes.

Their silence may not be incidental. It may be intentional. Perhaps even divine.

The absence of transitional forms isn't just a scientific dilemma—it's a philosophical one. What if these voids aren't failures of preservation, but clues of intentional craftsmanship? What if these gaps aren't failures of nature—but fingerprints of a Maker? One who designed life whole, distinct, and ready to thrive.

So the key question isn't just, *What's missing?*

It's this: *What is that silence saying?*

**Are we willing to listen to the stones that are crying out?**

Fossils aren't the only testimony in the case against evolution. If the rocks raise questions by what they don't show, the living world raises even more—by what it does show.

Beyond the gaps in the ground lies another kind of mystery—one built not on silence, but on striking similarities.

## Shared Design or Shared Descent?

Homology refers to the anatomical or genetic similarities across species—often cited as proof of common ancestry, especially in macroevolution.

But do shared features really prove evolution? Or could something else be at work?

Think of architect Frank Lloyd Wright. His buildings share similar elements—horizontal lines, open floor plans, harmony with nature. But that doesn't mean the houses evolved from one another. They reflect a common designer. Could biological similarities do the same—point to a single, intelligent Designer?

Consider the platypus—sometimes called "evolution's Frankenstein" because it combines traits from multiple animals. It has a duck's bill, beaver's tail, otter feet, lays eggs, and even produces venom. It sounds like a random mashup, but it's anything but dysfunctional. It hunts with electroreception, swims expertly, and thrives in its environment.

Rather than pointing at gradual evolution, the platypus defies it. Evolutionists struggle to place it on the evolutionary tree—it simply doesn't fit. Not a clumsy hybrid, but a precisely engineered creature. It doesn't shout "random mutation." It whispers "masterpiece."

These aren't the marks of evolutionary leftovers. They're signs of intentional design.

Taken with missing transitions, Darwin's theory looks less like settled science—and more like a framework on the verge of collapse.

The platypus isn't just a biological oddity. It's a vivid witness to design. Maybe the similarities across life aren't the echoes of evolution at all—but the unmistakable signature of a common Creator.

## Conclusion: The Fossil Record's Silent Verdict

Once regarded as evolution's strongest evidence, the story etched in stone now tells a different tale. The fossil record doesn't read like a seamless scroll—it reads like a disjointed archive, with missing pages where transitional forms should be.

Instead of a slow, continuous lineage, we see abrupt appearances and long stasis. Despite more than a century of digging—and the discovery of hundreds of millions of fossils—the "missing links" remain just that: missing.

The silence of the stones doesn't suggest delay. It points to a deeper truth: not evolution in progress, but creation revealed.

What does appear in the record? Creatures of astonishing complexity—right from the start. Functional anatomy. Integrated systems. Purposeful design. These aren't primitive prototypes—they arrive fully equipped for survival.

The coordination we see in living systems doesn't come from chaos—it bears the mark of intelligence.

As our understanding of genetics deepens, the case only grows stronger. DNA isn't just a molecule—it's code. Precise. Layered. Self-correcting. It doesn't resemble noise—it resembles language.

And information of that kind doesn't write itself by accident. It points to a mind behind the mechanism.

So maybe the silence in the fossil record isn't just an absence.

*Maybe it's a message.*

*Not an echo of blind chance—but a declaration of bold design.*

*The earth doesn't whisper evolution.*

*It declares engineering.*

And that brings us full circle to the question:

***What is that silence really saying?***

# The Cambrian Explosion

# Evolution's Big Bang

*Before the Cambrian, life was a bore—*
*Just slime in the soup, nothing to adore.*
*Then boom! New creatures—low to high—*
*Appeared so fast, and no one knows why.*
*It's like nature threw dice and rolled a surprise,*
*Leaving old theories rubbing their eyes.*

And that's not just poetic flair—it's exactly what the fossil record reveals.

If evolution were true, we'd expect life to evolve slowly and predictably—yet in one of Earth's oldest rock layers, something shocking occurs: complexity explodes into existence, without warning, without clear ancestors, and without explanation.

As we saw in the previous chapter, the fossil record doesn't support Darwin's vision of slow, step-by-step change.

But nowhere is that failure more dramatic than in the Cambrian layer.

## The Geologic Column: Evolution's Expected Story

To understand the challenge posed by the Cambrian Explosion, we first need to examine the geologic column.

From a secular viewpoint, the geologic column is a visual summary of Earth's rock layers, arranged from oldest at the bottom to the youngest at the top. These sedimentary layers supposedly formed slowly over vast spans of time, laid down by erosion, sedimentation, and tectonic shifts.

From a creationist perspective, the geologic column tells a different story—these layers were rapidly laid down during a global flood, sorted by density and particle size through hydraulic processes.

Evolutionary theory expects a gradual rise from simple to complex life, reflected as a clear step-by-step progression in the geologic column. The deepest layers, like the Precambrian, should contain only simple, single-celled organisms, while higher layers would show the slow emergence of more complex, multicellular life forms.

But this expectation breaks down in the Cambrian layer, which is directly above the Precambrian. Rather than a gradual buildup, we find a sudden burst of complex life forms—an abrupt leap in sophistication that completely defies evolutionary predictions.

## When Complexity Crashed the Party

The Cambrian Explosion presents a major challenge to Darwinian evolution. Across the globe, we find complex organisms in the Cambrian layer—fully formed, and without identifiable evolutionary ancestors. The Burgess Shale in Canada, one of the most famous Cambrian fossil beds, reveals a stunning variety of such creatures.

If life evolved slowly, we should see a clear trail of simpler forms in the Precambrian. Instead, the fossil record jumps from microbes to marvels—with no buildup in between. In the Cambrian layer, the stones don't just cry out—they proclaim together:

## Complexity was present from the very beginning.

Darwin himself predicted simpler life in older rocks, yet Cambrian strata showcase sophisticated body plans without clear predecessors.

Evolution expects a steady progression, but the Cambrian delivers a shock.

## "UH-OH... SOMETHING WRONG HERE!"

These Cambrian creatures aren't just anatomically complex—they imply a high level of genetic sophistication. Developing entirely new body plans would require new sets of genetic instructions. Yet there's no clear fossil or genetic evidence for such a gradual buildup, making the Cambrian's sudden complexity even harder to explain through evolutionary processes.

## Acknowledging the Paradox

Even evolutionary scientists recognize the Cambrian Explosion as a significant challenge to Darwin's model.

Dr. Jeffrey S. Levinton, professor of Ecology and Evolution at Stony Brook University, has called it evolution's "deepest paradox":

> *Why haven't new animal body plans continued to crawl out of the evolutionary cauldron during the past hundreds of millions of years? Why are the ancient body plans so stable?*[1]

Biologist Richard Dawkins also concedes:

> *We find many of them [Cambrian fossils] already in an advanced state of evolution, the very first time they appear.*[2]

These aren't statements from skeptics—they're admissions from committed evolutionists. The fossil record doesn't show a slow, steady march of transformation. It reveals a sudden leap in complexity—with no clear evolutionary trail behind it.

To evolutionists, this remains a scientific enigma. But from a creationist perspective, the pattern fits naturally. A global catastrophe—like the Flood described in Genesis—could explain both the rapid burial and exceptional preservation of diverse life forms, captured in the fossil record as if appearing suddenly and fully formed.

## Evolution's "Big Bang"

The Cambrian Explosion is often called "Evolution's Big Bang" because it contradicts the slow, step-by-step process Darwin envisioned. Within a narrow window of geologic time, a stunning variety of fully formed, complex organisms appear—most of which bear strong resemblance to modern species.

To account for this sudden appearance, paleontologist Stephen Jay Gould proposed the theory of *punctuated equilibrium*—the idea that evolution occurs in rapid bursts, separated by long periods of little or no change. Some have humorously referred to it as **"evolution on steroids."**

But the theory remains controversial—critics question what natural forces could drive such leaps in complexity. Evolution, we're told, moved at a crawl for billions of years—only to suddenly sprint, then stop again, without ever producing new foundational body plans.

This raises a serious question: Does punctuated equilibrium reflect a natural evolutionary process—or something that more closely resembles supernatural creation? Ironically, it sounds more like a miracle than a mechanism—precisely the kind of explanation many evolutionists aim to avoid.

## The Missing Fossils

Even Charles Darwin recognized the problem of the Cambrian Explosion:

> *To the question why we do not find rich fossiliferous deposits belonging to these assumed earliest periods prior to the Cambrian system, I can give no satisfactory answer.*[3]

Some evolutionists have argued that pre-Cambrian organisms were simply too soft-bodied to fossilize.

But this explanation falls well short—since we *do* have well-preserved fossils of delicate creatures like jellyfish and worms. What remains missing are the transitional forms leading up to the complex life of the Cambrian.

If evolution was gradual, we should find a trail of intermediate fossils bridging the gap from simple microbial life to advanced body plans.

Instead, the record leaps from single-celled simplicity to biological complexity—with no fossil roadmap in between.

The Cambrian Explosion stands as a dramatic leap, not a slow buildup—precisely the opposite of what Darwinian evolution predicts.

## The Flintstones Illusion: The Fossil Record Meets Bedrock

Imagine *The Flintstones*, where modern conveniences and Stone Age life exist side by side—foot-powered cars, dinosaur vacuum cleaners, and all. It's quirky and inconsistent by design.

The fossil record often appears just as disjointed. While evolution theory predicts a smooth, step-by-step progression from simple to complex life, what we actually observe is a series of abrupt appearances and unexplained gaps—much like the mixed-up world of Fred and Wilma.

To fill in these gaps, evolutionists often propose hypothetical "missing links." But after decades of digging, these links remain elusive. Just as *The Flintstones* blends incompatible elements to tell a story, evolutionary models frequently rely on imaginative reconstructions of the past—built more on what *should* be there than on what actually is.

Great for a cartoon. Not so great for scientific credibility.

The point is clear: the fossil record doesn't read like a continuous book—it looks more like scattered snapshots with missing pages. And the mystery isn't just what happened during the Cambrian—it's what *didn't* happen afterward.

## The Paradox of Extinction

Another puzzle in the fossil record is what happens *after* the Cambrian. While the Cambrian layer bursts with complex life, the diversity of species steadily declines in the layers above it. By the time we reach modern strata, most of those ancient creatures have vanished.

This trend doesn't support the idea of ongoing evolutionary advancement. If evolution were steadily driving innovation, we would expect to see an expanding variety of new species over time. In reality, the record shows widespread extinction—not continual emergence.

Paleontologist David Raup once estimated that roughly 99.9% of all species that have ever lived are now extinct.[4]

From a biblical creation perspective, this pattern makes sense: a sudden appearance of created life, followed by large-scale die-offs—consistent with a global catastrophe, such as the Flood described in Genesis.

If evolution were true, the fossil record should resemble a flourishing, ever-branching tree of life. Instead, it looks more like a collapsed forest—an initial explosion of growth, followed by sweeping loss.

## The Creationist View: A Sudden Creation of Life

The Cambrian Explosion aligns closely with the creation account in Genesis—a sudden appearance of fully formed life, "according to their kinds."

From this perspective, the complexity and diversity seen in Cambrian fossils are not the product of chance mutations, but of intentional design.

This phenomenon is not unique. Other biological systems—such as the human eye, the bacterial flagellum, and DNA itself—also exhibit levels of complexity that resist explanation by step-by-step evolutionary processes.

These examples, explored further in later chapters, reinforce the view that life was intentionally designed, not assembled by accident.

## The Biblical Perspective on Knowledge

Evolutionary theory rests on naturalistic assumptions—excluding the supernatural from the start. The Bible, by contrast, offers a different foundation for truth: one that begins with reverence for the Creator.

> *The fear of the Lord is the beginning of knowledge.*
> *— Proverbs 1:7*

Scripture presents a worldview that acknowledges God as the author and source of all life, offering the most coherent explanation for the sudden appearance of complex creatures in the fossil record. This perspective isn't new—Scripture warns that people will deliberately ignore the evidence of divine creation and global judgment in the past.

> *But they deliberately forget that long ago by God's word the heavens came into being and the earth was formed out of water and by water. By these waters also the world of that time was deluged and destroyed. — 2 Peter 3:5-6*

The apostle Peter's words remain strikingly relevant. Many continue to dismiss the reality of the global Flood and deny God's role in creation—yet the fossil record, particularly the Cambrian Explosion, speaks powerfully to both.

> *In Christ are hidden all the treasures of wisdom and knowledge. — Colossians 2:3*

A biblical worldview doesn't just describe what we see—it gives meaning to it. The sudden rise of complex life isn't merely a scientific anomaly. It's a declaration that life was created, not cobbled together by random chance.

> *My people are destroyed from lack of knowledge. — Hosea 4:6*

This verse serves as a warning: when people reject God's truth, confusion and distortion follow. The cultural consequences of evolutionary thinking will be explored in more detail in *Stumbling Block #14*.

## Conclusion

Darwin's theory stumbles in the face of the Cambrian Explosion—a burst of biological complexity too sudden, too complete, and too unexplained. Far from solving the mystery of life's origins, evolutionary theory often deepens it.

In contrast, the biblical creation model offers coherence, clarity, and purpose. The fossil record doesn't whisper evolution— *it shouts creation.*

# Deception in the Details

# When Hoaxes and Hype Shape Science

*Not all that glitters speaks the truth,*

*Some bones were carved to shape the proof.*

*When myths wear lab coats, crowds applaud—*

*But silence waits where truth is flawed.*

## When Fraud Becomes "Fact"

One of the most significant critiques of the theory of evolution is its long history of frauds, hoaxes, and misrepresentations— claims once celebrated as evidence, but later exposed as fabrications.

Eagerly promoted by the media and academia, these false "proofs" gave evolution an air of scientific certainty it hadn't truly earned.

Taken together, they ring powerful alarm bells—clear reminders of the dangers of uncritical acceptance in the name of science.

These episodes show how easily science can be hijacked by hype, and how difficult it is to correct course once a false narrative takes hold. Every so-called discovery became part of a story that few dared to challenge.

Here's a preview of some of the most notorious cases we'll examine—each one a story of hype, error, or outright deception:

- **Piltdown Man:** *A Man-Made Ape*

- **Archaeoraptor:** *The Piltdown Bird*

- **Nebraska Man:** One *Tooth, Big Mistake*

- **Lucy:** *The Poster Child of Evolution*

- **Neanderthal Man:** *Human All Along*

- **Vestigial Organs:** *Leftovers or Useful?*

- **Haeckel's Embryos:** *Drawn to Deceive*

- **Peppered Moths:** *Staged Science*

- **Miller-Urey Experiment:** *Chemistry or Creation?*

- **Scopes Trial:** *A Courtroom Drama*

We begin with one of the earliest—and most embarrassing—cases:

## Piltdown Man: A Man-Made Ape

In the early 20th century, a group of British scientists claimed they had uncovered the long-sought "missing link" between apes and humans.

The supposed "ape-man" fossil, named *Piltdown Man*, was heralded as a genuine evolutionary bridge between modern humans and our so-called knuckle-walking ancestors.

The scientific community and mainstream media responded with widespread, uncritical enthusiasm. Newspapers around the world celebrated the find. Peer-reviewed journals gave it legitimacy.

The excitement reached every corner of the academic world, and few dared to question it.

And for over 40 years, the academic establishment accepted this so-called evidence—many hoping it would finally silence creationist objections to evolution.

But *Piltdown Man* was a lie.

In 1953, the truth finally came out: Piltdown Man was an elaborate forgery—a composite of an orangutan's jawbone and a human skull, intentionally altered to look ancient. The bones had been stained to appear older, and the teeth filed down to mimic human wear patterns.

This wasn't an innocent error; it was meticulously planned and executed.

Historians now believe as many as six scientists colluded to mislead the global scientific community.

Even Webster's New World Dictionary, from 1988, acknowledges the fraud:

> *Piltdown Man was a supposed species of prehistoric human whose existence was presumed on the basis of skull fragments found in Piltdown (Sussex, England) about 1911 and exposed as a hoax in 1953.*[1]

It's unsettling to realize that for over four decades, some of the most respected voices in science and academia failed to question this false evidence.

Few imagined that something so widely accepted could be so completely fabricated.

What appeared in textbooks as fact helped shape evolutionary theory and misled generations. Only when the fraud was finally exposed did the full scope of the deception come to light—a carefully crafted ruse disguised as scientific discovery.

Once revealed, it was as if a monkey wrench had been thrown into the gears of evolution's propaganda machine.

The Piltdown hoax remains one of the most infamous scientific frauds in history. Tragically, it gave false credibility and artificial momentum to the theory of evolution—momentum it never truly earned.

Unfortunately, Piltdown Man wasn't an isolated deception. Over the years, other fabricated fossils and manipulated data have been used to shape public perception and reinforce evolutionary claims.

One of the most striking—and recent—examples? A feathered forgery now known as the "Piltdown Bird."

## Archaeoraptor: The Piltdown Bird

In 1997, a fossil that supposedly connected dinosaurs to birds was unearthed by a local Chinese farmer. This fossil was sent to the United States for testing and was named *Archaeoraptor*.

In 1999, *National Geographic* hailed it as the long-awaited "missing link" between dinosaurs and modern birds.

But it didn't take long for cracks to appear.

It turned out that *Archaeoraptor* was not one creature at all—it was a forgery. The fossil had been deliberately assembled by combining the tail of a small dinosaur with the upper body of an ancient bird.

What was celebrated as groundbreaking evidence for evolution quickly unraveled as a carefully constructed fake.

How did such an obvious fabrication escape scrutiny from respected scientists and editorial boards? In the rush to confirm evolutionary theory, careful scrutiny seemed to take a backseat.

As some critics observed, the fossil was accepted not because it passed rigorous examination, but because it fit the evolutionary story many in the scientific community were eager to confirm.

In this case, *Archaeoraptor* was more about "foul play" than legitimate science.

The late author and speaker Charles Colson exposed this troubling pattern in a compelling article, writing:

> *Most of us know National Geographic as the magazine we flip through at the doctor's office. Renowned for its stunning photography, National Geographic is one of the most highly esteemed periodicals in the world. That is, until last November's issue featured a discovery hailed as the best evidence to date for Darwin's so-called "missing link." But what was supposed to be startling news has turned out to be yet one more example of the scientific community peddling fraud as scientific fact.*
>
> *Then the truth came out. In reality, the Archaeoraptor fossil turned out to be the remains of two animals pieced together. While some call it an honest mistake, most now believe that it was actually an elaborate and deliberate act of fraudulent misrepresentation. But why, you may ask, is the scientific community so quick to embrace disreputable evidence? And why would an institution like National Geographic fail to take steps to confirm the reliability of such an amazing discovery?*

*The answer: They're desperate. You see, the lack of any evidence for transitional forms is one of Darwinism's dirty little secrets, and some scientists would do just about anything to keep it a secret – even to the point of fabricating evidence.[2]*

Within three years of Colson's eye-opening report, *National Geographic* officially recanted—because they had no choice. On November 20, 2002, they posted an article titled, *Dino Hoax Was Mainly Made of Ancient Bird, Study Says*. Here's an excerpt:

*The Archaeoraptor fossil was introduced in 1999 and hailed as the missing evolutionary link between carnivorous dinosaurs and modern birds. It was fairly quickly exposed as bogus, a composite containing the head and body of a primitive bird and the tail and hind limbs of a dromaeosaur dinosaur, glued together by a Chinese farmer.[3]*

ONE OF EVOLUTION'S
MANY FRAUDS

'archaeoraptor'

A Chinese
farmer glued the
front of a bird to the
back end of a reptile.

Once again, the zeal to prove Darwin right blinded the public. The *Archaeoraptor* scandal—now infamously dubbed the "Piltdown Bird"—echoes earlier evolutionary forgeries.

It's a sobering reminder that when belief drives science, fiction often gets dressed up as fact. Too often, scientists see what they want to see instead of examining what they've actually found.

The saga of misleading evidence doesn't end there. In fact, one of the most outrageous examples came from a single tooth—and it fooled the world.

## Nebraska Man: One Tooth, Big Mistake

The story of *Nebraska Man* is one of the most infamous scientific blunders of the 20th century—so outrageous, it's hard to decide whether to laugh or shake your head in disbelief.

In 1922, a single tooth was discovered in western Nebraska. Paleontologist Henry Fairfield Osborn, then president of the American Museum of Natural History, announced that the tooth belonged to a previously unknown species of early human. He claimed it showed characteristics that bridged the gap between apes and humans.

The media ran with the story. Without a single bone beyond that one tooth, artists created an entire prehistoric scene. A drawing in the *Illustrated London News* depicted *Nebraska Man* as a primitive, hairy, ape-like figure complete with his "wife" gathering food beside him. It was, quite literally, a completely fabricated family portrait based on one molar.

As Dr. Duane Gish later commented in *Evolution: The Fossils Still Say No*, ***"A scientist made a man out of a pig—and the pig made a monkey out of the scientist!"***[4]

By 1927, the truth came to light. Additional fossils found in the same area confirmed that the tooth wasn't human at all—it belonged to an extinct species of peccary, a pig-like mammal common in North America. *Nebraska Man* vanished from textbooks as quickly as it appeared.

This embarrassing episode revealed just how easily speculation can masquerade as science when it fits a favored narrative. The desire to find a "missing link" led to a leap of imagination rather than a foundation of evidence. It's a wake-up call: when interpretation outruns the data, science can slip into ideology in disguise.

Yet despite these high-profile blunders, the hunt for evolutionary fossil icons continued. And one of the most famous—and most controversial—figures to emerge from this search was a fossil nicknamed *Lucy*, often promoted as the definitive human ancestor.

## Lucy: The Poster Child of Evolution

Discovered in 1974 by paleoanthropologist Donald Johanson in Ethiopia, *Lucy*—scientifically named *Australopithecus afarensis*—was quickly hailed as the much-anticipated "missing link" between apes and humans. Her partial skeleton became a media sensation. Museums built life-size models. Textbooks revised timelines.

Yet beneath the hype, Lucy's fossil was fragmentary—especially the skull—and her appearance was reconstructed with a heavy dose of imagination. As renowned paleoanthropologist Richard Leakey admitted in 1983:

> *Lucy's skull is so incomplete that most of it is imagination made of plaster of Paris.*[5]

Using advanced analysis, Dr. Charles Oxnard concluded that australopithecines, the group that includes Lucy, differ *more from both apes and humans than apes and humans differ from each other.*[6]

In other words, Lucy was likely not a human ancestor at all—but a distinct species. Further anatomical studies raised even more questions. While some bones suggest upright walking, others—like her cone-shaped rib cage—exhibit traits seen in tree-dwelling apes. Even the famous Laetoli footprints, often linked to her, may have been made by a different, more advanced hominin.

Still, Lucy remains in school curricula and museum displays as a cornerstone of human evolution. Why? Because once a fossil becomes a symbol, it's hard to let go—even when the science moves on. Lucy's story highlights a deeper issue: when evolutionary claims outpace the evidence, science turns into storytelling.

Lucy's discovery, while groundbreaking in popular culture, highlights the ongoing debate and uncertainty within paleoanthropology. Far from a clear "missing link," Lucy's incomplete fossil record and contested interpretations illustrate the challenges of definitively tracing human ancestry. This case underscores the need for careful evaluation, not quick acceptance, of evolutionary narratives.

Next, we examine another iconic figure in human evolution—Neanderthal Man—and how modern science has reshaped our understanding of this ancient cousin.

## Neanderthal Man: *Human All Along*

For decades, *Neanderthal Man* was portrayed as a stooped, grunting brute—part ape, part man—used in textbooks and museums to suggest a transitional form in human evolution. This image became iconic, reinforcing the idea of a slow ascent from beast to modern human.

But as technology advanced, that caricature began to crumble. 3D scanning and digital reconstruction gave scientists new insight into Neanderthal anatomy—from brain folds to cranial capacity—and what they discovered was unexpected: *Neanderthals were strikingly human*. Their brains were within modern size ranges, and their robust features reflected adaptation to cold climates—not a lack of intelligence. Their tools, burial practices, and even musical instruments point to abstract thinking, language, and culture. As paleoanthropologist Dr. Erik Trinkaus noted, Neanderthals were "fully human beings." Their differences were environmental, not evolutionary.[7]

The creationist perspective finds no conflict here. Neanderthals were human from the beginning. Their features reflect diversity within mankind—not descent from apes.

The evidence aligns with the Genesis account: *God created both Adam and apes, but He did not create Adam from apes.*

## "OBVIOUSLY THIS IS THE SKULL OF A MONKEY!"

Though evolutionary theory has adjusted over time, it still clings to a narrative of upward progress. But *Neanderthal Man* tells a different story—not of primitive to advanced, but of mistaken assumptions corrected by better science.

The so-called caveman was no beast. *He was fully human.*

Next, we'll look at how "vestigial organs" have been misused to back evolutionary claims.

## Vestigial Organs — Leftovers or Useful?

For decades, evolutionists claimed certain human body parts—like the appendix, male nipples, tailbone, and goosebumps—were "vestigial organs," supposedly useless leftovers from our evolutionary past. If evolution were true, we would expect many more of these so-called remnants.

This textbook narrative implied imperfections in design, subtly arguing against a Creator. But modern science has steadily overturned that claim. Today, nearly all so-called vestigial organs serve real, important functions.

The appendix, once thought to be a useless relic, actually helps regulate immune responses and supports healthy gut bacteria. The tailbone anchors key pelvic muscles. Goosebumps help regulate body temperature and skin oil. Male nipples arise from efficient embryonic development and have cosmetic or sensory roles.

What were once called evolutionary leftovers are now understood as functional, integrated parts of the human body. Far from flawed remnants, they reflect intentional, intelligent design— a truth long obscured by evolutionary bias.

If vestigial organs were a case of misinterpretation, the story of Haeckel's embryos was outright fraud.

## Haeckel's Embryos: *Drawn to Deceive*

In 1868, German biologist Ernst Haeckel published embryo drawings claiming that early-stage embryos of vertebrates—that is, animals with backbones, including humans—looked nearly identical.

He presented this as visual proof of evolution, insisting that embryos replay their species' evolutionary history.

*The problem? The drawings were fake.*

Haeckel had exaggerated similarities, altered features, and even duplicated images to fit Darwin's theory.

As embryology advanced, scientists exposed his work as fraudulent—yet the images still appeared in textbooks for decades.

Some defended their use as educational tools—even after acknowledging they were inaccurate. But promoting false evidence doesn't teach science—it distorts it.

Haeckel's embryos remain a stark reminder: when ideology drives science, truth becomes the first casualty.

The damage wasn't just historical—it shaped how generations would view biology and origins.

And the tactics weren't limited to drawings. Even staged experiments—like the case of the peppered moths—were used to prop up evolutionary claims.

## Peppered Moths: *Staged Science*

The peppered moth experiment is often hailed as a textbook example of natural selection. In polluted areas, dark-colored moths supposedly blended into soot-covered trees, avoiding predators and surviving at higher rates. It's long been used to support Darwin's theory.

But the experiment was flawed. Researchers pinned moths to tree trunks—not where moths naturally rest—to stage photos that appeared to confirm the theory.

These artificial setups created an illusion of camouflage and skewed the results.

In reality, moths rest under branches or in sheltered areas, not out in the open on tree trunks. This critical detail undermined the entire premise.

Later studies exposed the experiment's manipulation and oversights. Rather than proving evolution, the peppered moth story relied on staged photos, selective reporting, and an incomplete view of moth behavior.

Just as this narrative unraveled, another famous evolutionary "proof"—this time about the origin of life—would face its own scientific reckoning.

## Miller-Urey Experiment: *Chemistry or Creation?*

The 1953 Miller-Urey experiment aimed to show how life could arise from non-living matter—supporting Darwinian evolution. By sending electrical sparks through a mix of gases thought to mimic early Earth, the experiment produced amino acids, the building blocks of proteins. This was initially hailed as proof that life could form spontaneously.

But the celebration was premature.

Later research revealed that the gases used—methane and ammonia—likely didn't reflect Earth's actual early atmosphere, making the results irrelevant to real conditions.

More importantly, forming amino acids is only a tiny step toward life. The experiment never produced anything close to a living cell. If trained scientists, under precise conditions, couldn't produce life from non-life, how could random chance do it in a chaotic, uncontrolled environment?

The experiment's failure to generate life highlights just how wide the gap is between molecules and living organisms. Rather than proving evolution, it points to the possibility of intelligent design—a purposeful Creator, not accidental chemistry.

The Miller-Urey experiment remains a powerful example of how early scientific claims in support of evolution often crumble under closer scrutiny.

But the push to promote evolution didn't stop in the lab—it spilled into the courtroom and headlines, shaping public opinion through drama and deception. One of the most famous cases? A staged legal battle that came to define the culture war over evolution in American schools.

***Disclaimer: Monkeys, please wait until fully human before proceeding.***

## The Scopes Trial: A Courtroom Drama

The 1925 Scopes "Monkey" Trial is often portrayed as a dramatic showdown between science and religion. In reality, it was a staged event built on false pretenses.

John Scopes, a high school teacher in Dayton, Tennessee, was accused of violating the Butler Act, which banned teaching evolution in public schools. But Scopes was no rogue rebel—he was recruited by local leaders and the ACLU to provoke a test case for publicity. He hadn't even taught evolution; he merely agreed to participate.

The trial became a media circus. With celebrity lawyers Clarence Darrow and William Jennings Bryan facing off, national attention turned the courtroom into a platform for ideology.

Darrow, an agnostic, used the stage to attack religion, while Bryan defended biblical creation—not through scientific debate, but moral conviction.

The trial wasn't about science. It was pure spectacle.

- **Orchestrated Case**: Scopes was a reluctant figurehead, chosen to trigger legal drama—not because he defied the law, but because Dayton needed attention.

- **Media Manipulation**: National outlets framed the trial as enlightened science vs. backward religion, ignoring the local nuances and turning Bryan into a caricature.

- **No Scientific Discussion**: The trial devolved into personal attacks and theological debate. Evolution's scientific merits were barely addressed.

Scopes was found guilty and fined $100—a conviction later overturned— but the trial's real legacy was cultural. It cemented a false narrative: that science triumphed over superstition. In truth, it was a publicity stunt wrapped in ideology, not a fair or factual debate.

The *Scopes Trial* is a lasting example of how public perception can be shaped—not by evidence—but by agendas, media spin, and cultural pressure.

## Conclusion: The Deceptive Power of Unverified Claims

The *Scopes Trial*, along with the other examples explored in this chapter, reveals how media, politics, and selective storytelling can shape public opinion—often at the expense of truth.

Rather than fostering open scientific inquiry, these events became platforms for advancing ideological agendas, leaving lasting impacts on how evolution is taught in schools today.

Fabrications like *Piltdown Man* and manipulated studies like the *peppered moth* experiment expose a troubling pattern: deception used to promote a worldview that excludes the possibility of a Creator. These cases underscore a critical lesson: when claims go unchallenged, falsehoods gain power.

Too often, conclusions are guided more by cultural momentum and political pressure than by objective scientific evidence.

As we continue to explore the origins of life and humanity, we must stay discerning. What's often presented as fact may rest on shaky ground. Textbooks and media narratives can mislead when they go unchallenged.

By pursuing truth based on verified evidence—not ideology—we guard our understanding from being shaped by agenda rather than reality.

# Dinosaurs A Fresh Look

*From legends of dragons to footprints in stone,*

*Strange signs suggest we weren't alone—*

*What if the timeline's not etched in rock,*

*Could dinosaurs and man have shared a walk?*

## The Big Question: Did Humans and Dinosaurs Coexist?

For decades, science has taught that dinosaurs died out 65 million years before humans appeared.

Yet this widely accepted timeline raises some intriguing questions: Could it be incomplete, or even misinterpreted?

Across diverse cultures, dragon legends—ranging from fire-breathers in European folklore to serpent-like creatures in China—share surprising similarities with dinosaur fossils.

Could these pervasive myths reflect distorted memories of real encounters between early humans and dinosaurs?

Might these legends be more than myth—echoes of forgotten history?

## The Dragon Connection: Myth or Memory?

Although dinosaurs are considered extinct, creatures like the modern Komodo dragon—a large, powerful lizard with thick scales and a menacing presence—share striking traits with prehistoric reptiles.[1]

This resemblance has led some to suggest that ancient "dragon" legends may reflect real encounters with dinosaur-like creatures.

It could help explain why cultures across the globe—who had no contact with each other—described dragons in ways that closely match known dinosaur anatomy.

Such global consistency raises a compelling question: were these stories invented, or do they reflect real memories?

Dr. Leland Niermann notes in the *Creation Ex Nihilo Technical Journal*:

> *The oldest record of possible dinosaur bones is in a Chinese book written between A.D. 265 and 317. It mentions 'dragon bones' found at Wucheng, in Sichuan Province... To the Chinese, 'Dragon bones' and 'Dinosaur bones' were one and the same.*[2]

Perhaps "dragon" wasn't always just a mythical label. It may have been the ancient world's name for something very real that once roamed the Earth.

## Global Signs of Human-Dinosaur Encounters

If humans and dinosaurs once lived side by side, we would expect to find stories, symbols, or depictions of them across ancient cultures. And that's exactly what we see. From Chinese "dragon bones" and European dragon legends to African cave paintings and Native American tales of reptilian beasts, accounts of dragon-like creatures appear on nearly every continent. The similarity of these ancient accounts across continents may reflect a common reality remembered differently by diverse cultures.

In a very unusual attempt to explain this phenomenon, astronomer Carl Sagan proposed that dragon legends might stem from inherited "fossil memories"—genetic impressions passed down from early mammals.[3] But the idea that ancient fears of predators were somehow encoded into DNA and then expressed as consistent dragon myths across the world strains all credibility.

Could there be a far simpler explanation? These widespread legends may be based on real encounters with now-extinct creatures—creatures we today would call dinosaurs.

## Dinosaurs in the Bible: Hidden Records?

The Bible may offer more clues about dinosaurs than many realize. In the Book of Job, God describes a powerful, towering creature called *Behemoth*:

*Look at Behemoth, which I made along with you and which feeds on grass like an ox. What strength it has in its loins, what power in the muscles of its belly! Its tail sways like a cedar; the sinews of its thighs are close-knit. Its bones are tubes of bronze, its limbs like rods of iron. — Job 40:15–18*

Some scholars argue that this creature more closely resembles a sauropod dinosaur than any living animal. And it's worth noting: the word "dinosaur" wasn't coined until 1842. Before that, ancient people likely referred to such creatures using names like dragons, Behemoth, or terrible lizards.

Beyond the Bible, ancient Egyptian hieroglyphs and early artwork from other civilizations frequently depict large, reptilian creatures.

These consistent images across cultures strongly suggest that people in the distant past may have had real encounters with animals remarkably similar to what we now call dinosaurs.

## Dinosaur Depictions in Ancient Art

Some researchers point to visual evidence of coexistence. In Zimbabwe's Gorozomzi Hills, for instance, cave paintings dated to around 1500 B.C. appear to depict a long-necked, four-legged creature that closely resembles a brontosaurus.

As the *Los Angeles Herald Examiner* reported:

*A fantastic mystery has developed over a set of cave paintings found in the Gorozomzi Hills, 25 miles from Salisbury [in Rhodesia, now Zimbabwe]. For the paintings include a brontosaurus—the 67-foot, 30-ton creature scientists believed became extinct millions of years before man appeared on Earth.[4]*

These images were created by the Bushmen, a group well known for their accurate representations of local wildlife. This raises a challenging question: Why are dinosaur-like figures so often dismissed—especially when other animals in the same paintings, like giraffes and elephants, are accepted as authentic depictions of real creatures?

Similar depictions appear around the world—such as dinosaur-like carvings at Cambodia's Ta Prohm temple or petroglyphs in South America—suggesting a broader pattern worth investigating.

## Dinosaurs in Global Ancient Texts

Numerous ancient texts from diverse cultures describe creatures that are remarkably similar to dinosaurs:

- **Greek historian Herodotus** mentions massive serpentine reptiles.

- **Han Dynasty texts** describe "dragon bones" used for medicine.

- **The Epic of Gilgamesh** includes dragon-like beasts encountered by humans.

Global legends—such as the European story of Saint George slaying a dragon or the Amaru serpent myths in South America's Andean cultures—depict powerful, often reptilian creatures. These stories could reflect ancient sightings of large dinosaurs or pterosaurs.

The consistency of these worldwide accounts suggests a shared historical experience with such creatures, rather than mere cultural coincidence or independent mythical invention.

## Cultural Amnesia or Evidence Suppression

Despite this accumulating body of suggestive evidence—from ancient texts and artwork to global legends—mainstream science consistently rejects the idea of human-dinosaur coexistence. Some critics argue this stance reflects not just scientific skepticism, but a deeper reluctance to entertain evidence that might undermine the prevailing evolutionary model.

Dr. Andrew Snelling, a geologist and creation scientist, has openly expressed concern over the suppression of alternative interpretations of geological data:

> I have tried repeatedly to submit evidence that
> contradicts the present evolutionary geological column,
> only to have it outright suppressed or ignored by

*journal editors and university reviewers. It seems that
anything suggesting the fossil record can be explained
by global Flood processes must be quashed if
mainstream acceptance is to be maintained.*[5]

He further elaborates:

*If, as many creationist geologists believe, the majority
of the geological column represents flood sediments
and post-flood geophysical activity, then the mammoth,
dinosaur and all humans existed simultaneously.*[6]

Such staunch resistance, as observed by Snelling and others,
may indeed reflect a cultural commitment to evolutionary
orthodoxy—sometimes at the expense of objective scientific
inquiry.

## Questioning Deep Time: Scientific Challenges

One significant challenge to deep-time assumptions comes
from the unexpected presence of carbon-14 in some dinosaur
fossils, as well as in coal and diamonds.

Since carbon-14 decays rapidly with a half-life of 5,730 years,
it should be undetectable after approximately 50,000 years.

Yet, measurable amounts have reportedly been found in over
fifty fossils that are conventionally dated at tens of millions of
years.[7]

These findings are controversial. Mainstream scientists often dismiss them as contamination, suggesting modern carbon may have infiltrated the samples through groundwater, microbial activity, or handling.

However, critics of the deep-time model argue that the consistent detection of C-14 across numerous samples and independent labs points to something more than isolated contamination.

At a minimum, these discoveries call for thorough independent investigation and highlight the crucial need for open scientific dialogue regarding the fundamental assumptions embedded in radiometric dating methods.

## Geological Evidence for a Young Earth

Various geological and fossil evidences support the possibility of a much younger Earth and more rapid processes:

- **Burgess Shale:** The exquisite preservation of delicate, soft-bodied organisms in the Burgess Shale suggests rapid burial under fine sediment. Such remarkable detail indicates sudden, high-energy conditions, rather than slow accumulation over millions of years, which would typically lead to decomposition.

- **Mount St. Helens (1980) Eruption:** This catastrophic event vividly demonstrated how volcanic activity can rapidly deposit thick layers of sediment. Within days, massive mudflows carved out canyons and created stratified formations, remarkably mimicking features that conventional geology typically assumes took millennia to develop.

- **The "Little Grand Canyon":** Formed in days by a massive mudflow at Mount St. Helens, this feature has sharply defined sediment layers and steep walls, resembling the real Grand Canyon.

These examples support the idea that dinosaur fossil beds may have formed during large-scale catastrophes—like a global flood—rather than slow, uniform geological processes.

Even some secular scientists now acknowledge that geological features can form far more rapidly than once believed, a concept explored further in Stumbling Blocks #11: *The Case for a Global Flood.*

## The Genesis Perspective: A Shorter Timeline

From a biblical perspective, dinosaurs fit naturally within the Genesis creation account. Scripture teaches that God created the world and all land animals—including dinosaurs—as well as humans on Day Six, placing the events of creation around 6,000 to

7,000 years ago. This sharply contrasts with the mainstream scientific view of a 4.5 billion-year-old Earth.

Much of the fossil record—especially the evidence for sudden burial and preservation—aligns more closely with the biblical account of a global flood.

According to Genesis, a cataclysmic deluge in Noah's day reshaped Earth's surface and rapidly buried countless organisms, including many dinosaurs.

Rather than a slow process spanning millions of years, this kind of event would have created ideal fossil-forming conditions: vast amounts of mud, sediment, and water quickly and completely covering plants and animals.

This could explain why so many dinosaur fossils are found in massive "fossil graveyards," often with skeletons twisted backward or in classic drowning postures. These fossil beds appear worldwide—even in places like Antarctica. Well-known sites such as the *Hell Creek Formation* in Montana and the *Morrison Formation* in the western United States have yielded enormous numbers of dinosaur bones, often jumbled together in chaotic layers—suggesting sudden, catastrophic burial.

Similar fossil concentrations in the *Lourinhã Formation* of Portugal and the *Nemegt Basin* in Mongolia also point to mass-death events rather than slow, steady deposition over time.

The rock layers containing dinosaur fossils frequently show signs of rapid, high-energy processes, not gradual accumulation.

Some dinosaurs may have survived briefly after the Flood, only to go extinct in the harsher post-Flood world—due to climate shifts, reduced food sources, or possibly human hunting.

## Soft Tissue Discoveries: A Game-Changer?

While ancient accounts and geological clues provide compelling reasons to question the conventional timeline, perhaps the most visceral and undeniable challenge comes not from interpretations of texts or rocks, but from the very makeup of the fossils themselves.

The scientific world was stunned by the discovery of soft tissue in dinosaur fossils—so remarkably preserved that they still contained skin, muscle fibers, and blood vessels.[8]

Further research into these extraordinary finds has even yielded evidence of proteins, chromosomes, and chemical markers of DNA, challenging long-held assumptions about fossilization.[9]

These aren't isolated anomalies. Similar discoveries have now been made in a growing number of dinosaur species and formations, including *Triceratops, Archaeopteryx, and Hadrosaurs.*

The repeated discovery of short-lived biological molecules—such as collagen, actin, myosin, and hemoglobin—along with intact blood vessels and exceptionally preserved bone cells (complete with delicate filopodia) profoundly challenges the idea that these fossils are millions of years old. (*Filopodia are slender, finger-like projections that help cells sense their environment.*)

In 2007, *National Geographic* reported the discovery of "Dakota," a fossilized dinosaur with remarkably intact skin, muscle, and blood vessels:

> *Scientists today announced the discovery of an extraordinarily preserved 'dinosaur mummy' with much of its tissues and bones still encased in an uncollapsed envelope of skin.*[10]

In other words, the dinosaur's skin had not collapsed or decayed—something considered virtually impossible if it were truly millions of years old.

Since organic tissue breaks down rapidly, the presence of soft tissue in dozens of dinosaur fossils defies conventional expectations. This strong evidence suggests these remains may be far younger than widely assumed.

Further discoveries followed. In 2009, another *National Geographic* article described blood vessels and connective tissues in an 80-million-year-old duck-billed dinosaur:

*The fossilized leg of an 80-million-year-old duck-billed dinosaur has yielded the oldest known soft-tissue proteins—including blood vessels and other connective tissue, and likely blood cell proteins.*[11]

It's difficult to fathom how such delicate biological materials could survive 80 million years.

Soft tissues and proteins are typically the first parts of an organism to decompose—yet in many fossils, they remain astonishingly intact.

Paleontologist Dr. Mary Schweitzer has proposed a mechanism known as *iron preservation* to explain the survival of soft tissue in dinosaur fossils.

Her hypothesis suggests that iron from hemoglobin in dinosaur blood may have acted as a preservative by promoting cross-linking in proteins—similar to how tanning preserves leather.[12]

Schweitzer's lab demonstrated that iron-rich solutions can slow tissue decay in the short term.

However, these experiments only demonstrated preservation over days to months—not millions of years.

Moreover, iron is also known to accelerate oxidative damage over time, which would typically lead to further degradation—not preservation.

Additionally, some researchers have proposed *microbial biofilms*—bacterial residue that can sometimes mimic tissue shapes—as an alternative explanation.[13]

But these films have failed to replicate the detailed cellular structures and preserved proteins found in many fossils.

Given these limitations, both the iron and biofilm hypotheses remain highly speculative and controversial, falling short of explaining the remarkable and repeated discovery of well-preserved soft tissues in dinosaur fossils.

The ongoing discovery of fragile biological materials in dinosaur remains poses a significant challenge to the conventional timescale.

If these fossils are far younger than assumed, could they date back thousands—not millions—of years?

## Conclusion: Rethinking Earth's Timeline

When evidence from multiple disciplines converges, it demands serious attention. Taken as a whole, the diverse body of evidence—from pervasive ancient stories and artwork to remarkably preserved dinosaur tissue discoveries and compelling examples of rapid geological change—strongly suggests a possible overlap between humans and dinosaurs.

If the Bible's creation account is accurate, this overlap makes perfect sense. Dinosaurs, from this perspective, would have been created alongside humans, lived on the Earth until the global Flood, and possibly survived for a short time in the post-Flood world.

As for how dinosaurs could have fit on Noah's Ark, the answer is surprisingly straightforward: God likely instructed Noah to bring **juvenile dinosaurs—smaller, lighter, and more manageable**, making them ideal passengers for the journey.

As our understanding of Earth's past continues to evolve, so too must our willingness to question long-held assumptions. When viewed together, the case for human-dinosaur coexistence is not only compelling—*it's becoming increasingly difficult to dismiss.*

# The Mousetrap Challenge

# Do All the Parts Matter?

*A trap with no spring won't catch a thing—*

*It's either all there, or it won't do a thing,*

*So when nature builds with parts that must stay,*

*Could random chance really work that way?*

## A Broken Blueprint

Jake ordered a drone kit online—"easy assembly," the ad promised. No instructions? No problem. He'd built IKEA furniture before.

He snapped together the frame, then attached the rotor arms—forgetting the screws. He wired up the camera before adding the power system. It looked slick—until it slid off the table and cracked.

Next came the rotors—no propellers. He powered it up. The drone buzzed like an angry wasp and toppled over. He connected the controller but forgot the receiver. He added GPS, but the drone had no idea where it was. Finally, he installed the battery. The drone malfunctioned, sparking and emitting smoke as it powered up.

Three hours later, Jake was left with a damaged coffee table, a few bandages, and a pile of unusable parts.

The drone had nearly everything—but it didn't *almost* fly.

It didn't fly at all.

Unless all the parts are aligned, assembled, and working in harmony, flight isn't even possible.

Jake's failed drone wasn't just a technical mishap—it illustrates a profound truth: complex mechanisms only work when all the right parts are present and functional, all at once.

If even one part is missing or out of sync, the system fails entirely.

This all-or-nothing principle is key to understanding a concept called *irreducible complexity*.

## What Bacteria Can Teach Us

Just like Jake's drone, many biological systems require every part to be in place and fully functional from the start.

The absence of even one component can render the entire system nonfunctional.

This challenges the idea that complex organisms can gradually evolve through small, random changes.

This is exactly what biochemist Michael Behe identified when he coined the term irreducible complexity:

> *Irreducibly complex structures are made up of several well-matched, interacting parts that work together to perform a basic function. If any one part is removed, the system ceases to function altogether.*[1]

To explain this, Behe uses a simple analogy: a **mousetrap**. Any standard mousetrap needs several components:

- Spring
- Trigger
- Catch
- Platform
- Hammer

Take away any one of those, and it stops being a trap—it becomes junk.

The same principle applies in biology. Many systems—far more intricate than a mousetrap—simply cannot function unless all their parts are present and arranged correctly.

Consider, for example, the bacterial flagellum.

## Nature's Microscopic Machines

The bacterial flagellum—highlighted in the 2002 documentary *Unlocking the Mystery of Life*—is a truly awe-inspiring example of microscopic engineering.

> *Through the microscope, we observe the E. coli bacterial flagellum. The bacterial flagellum is what propels E. coli bacteria through its microscopic world. It consists of about 40 individual protein parts, including a stator, rotor, drive-shaft, U-joint, and propeller. It's a microscopic outboard motor! The individual parts come into focus when magnified 50,000 times (using electron micrographs). And even though these microscopic outboard motors run at an incredible 100,000 rpm, they can stop on a microscopic dime. It takes only a quarter turn for them to stop, shift directions and start spinning 100,000 rpm in the opposite direction! The flagella motor has two gears (forward and reverse), is water-cooled, and is hardwired into a signal transduction (sensory mechanism) so that it receives feedback from its environment.[2]*

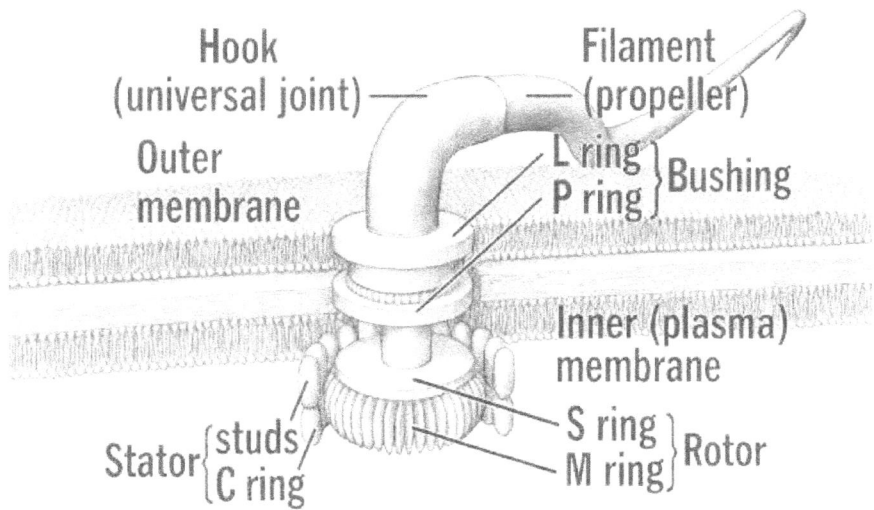

In short, the bacterial flagellum is a **nanoscopic outboard motor**—a marvel of engineering that rivals (or exceeds) anything humans have ever built.

It isn't "almost" a motor—it's a complete, fully operational machine.

So here's the question:

How could such a precisely engineered, high-speed, multi-component system come into existence one piece at a time—by random mutation and natural selection?

It couldn't.

Just like Jake's drone or a mousetrap, it's all-or-nothing.

## The Problem with Piecemeal Progress

Darwinian evolution argues that life's complexity arose gradually—through small, step-by-step modifications across vast spans of time.

But irreducibly complex structures like the bacterial flagellum challenge that idea head-on.

Why? Because these systems don't work in parts.

You can't have an "almost" working flagellum, any more than you can have a "mostly" built mousetrap or an "incomplete" drone that still flies.

It either works—or it doesn't.

## The Type III Secretion System Argument

Some evolutionary biologists attempt to explain the flagellum by pointing to a structure called the Type III Secretion System (TTSS)—a tiny molecular syringe that some bacteria use to inject toxins into host cells.

It shares a handful of protein components with the flagellum, so the theory goes: *perhaps the flagellum evolved from the TTSS— or vice versa.*

But this claim doesn't hold up under closer scrutiny.

William Dembski explains:

*Finding a subsystem within a functional system that performs some other function is hardly an argument for the original system evolving from that other system. One might just as well say that because the motor of a motorcycle can be used as a blender, therefore the motor evolved into the motorcycle.*[3]

Both the TTSS and the flagellum are complex, multi-part systems—but they're not interchangeable. One injects toxins. The other propels. And there's no known evolutionary pathway that shows how one became the other through gradual modifications. Having a few biological parts doesn't mean you're halfway to a motor—you might just have a pile of useless proteins.

Some scientists suggest *exaptation*—the idea that parts evolved for one use and were later repurposed for another. But this doesn't solve the problem. A flagellum isn't just a pile of useful parts—it's a tightly coordinated machine. It's not enough to have components lying around; they must be assembled and function together in perfect sync from the start.

## Why Gradual Evolution Falls Short

The core problem with gradual evolution is that systems like the bacterial flagellum simply don't allow for intermediate stages. There's no "almost working" version—it either functions fully or not at all. Think of a mousetrap: If even one part is missing, it won't catch a mouse. Similarly, the flagellum requires every single component to be present and working in harmony from the very start.

If gradual evolution were true, transitional forms should be abundant—not rare. The natural world ought to be teeming with organisms stuck in evolutionary limbo: creatures developing wings that don't yet allow flight, eyes that don't yet see, or lungs that don't yet breathe. These incomplete, unfinished systems would be the norm rather than the exception.

Yet, we don't observe this—neither in living species nor in the fossil record. Instead, most organisms appear suddenly, fully formed, and functionally complete. This striking absence of transitional forms poses a serious challenge to the story of gradual evolution.

Irreducibly complex arrangements like the flagellum—or many human biological systems—cannot tolerate incomplete versions of themselves.

Consider the human body:

If evolution by gradual steps were true, how did humans survive while their vital organs were supposedly in constant states of development? Take these systems, for example, and consider the difficulty in explaining how each could evolve independently without the others fully functioning:

- **Optical system**: What came first—the desire to see or the ability to see? Which component evolved first: the pupil, lens, retina, optic nerve, cornea, iris, veins, rods, cones, tear ducts, or the highly advanced brain that not only receives but interprets visual information?

- **Respiratory system**: What evolved first—the lungs, mucus lining, windpipe, or the precisely calibrated mixture of gases required to sustain life?

- **Circulatory system**: What came first—the blood, plasma, heart, platelets, veins, arteries, capillaries, valves, or the complex clotting mechanism, where a failure of a single step can be fatal?

Like the mousetrap, these systems are fully functional wholes—not cobbled-together inventions waiting millions of years to start working. The human eye consists of about 40 distinct, interconnected subsystems that must work in precise coordination to function. These parts had to appear together all at once—or we simply wouldn't be able to see.

## The Impossibility of Life's Random Assembly

The question remains: How could the many intricate components of irreducibly complex biological machinery come together by mere happenstance?

The bacterial flagellum's parts must align with absolute precision to function—yet randomness alone cannot produce such order.

The complexity of life demands far more than blind chance; it requires a blueprint and an architect who orchestrates the formation and assembly of every component in every organism.

Charles Darwin himself acknowledged this challenge in *The Origin of Species.* He stated that his theory would be disproved if it could be shown that any organ—such as the eye, ear, brain, or other highly complex systems—could not have evolved through gradual, successive modifications. As Darwin wrote:

> *If it could be demonstrated that any complex organ existed, which could not possibly have been formed by numerous, successive, slight modifications, my theory would absolutely break down.*[4]

The bacterial flagellum stands as a perfect example of such a complex system that challenges Darwin's theory. Like all irreducibly complex mechanism, its parts must be fully formed and work together right from the start.

Evolution by random, gradual processes simply cannot explain how these systems arise without the guiding hand of an intelligent designer.

## The Threat to Academic Freedom

As we've seen with the bacterial flagellum and other irreducibly complex systems, evolutionary theory is far from being airtight.

Yet today, in the scientific community, researchers who question this theory often face significant professional risks.

As highlighted in Ben Stein's documentary *Expelled: No Intelligence Allowed*, many scholars argue that academic freedom is under threat for those who question the dominant evolutionary narrative.[5]

Genuine science thrives on the free exchange of ideas. When alternative views—such as Intelligent Design—are silenced or suppressed, it raises concerns that the theory of evolution may not be as robust as it is often portrayed.

If evolutionary theory were truly irrefutable, it should welcome scrutiny. The fact that it doesn't raises serious questions about what's really being protected—scientific truth or a preferred worldview?

## An Answer That Shapes Everything

Just like Jake's drone or a simple mousetrap, life's most intricate systems must be fully integrated. They can't function when assembled piecemeal. Every part must be in the right place, operating in perfect harmony from the start.

Think of the bacterial flagellum, the human eye, or the blood-clotting mechanism—these aren't products of chance. They're integrated, finely tuned systems, operating with precision unmatched by any human invention.

They don't merely suggest complexity—they reveal coordination, intention, and purposeful design.

Random mutations and natural selection don't build feedback loops, build propulsion motors, or calibrate chemical pathways with surgical precision.

But purposeful design can.

## Why It Matters

Behind the science lies a deeper question—one of worldview: Is life the result of blind, random forces—or the deliberate craftsmanship of an intelligent Designer?

This isn't just about how life began. It's about what life means.

Your answer doesn't just explain the past; it shapes purpose, meaning, and the future—a topic we'll explore further in the final chapter.

# Do Evolution's Mechanisms Deliver?

*Mutations are random, they change the game,*

*But most lead to harm, not evolutionary fame.*

## Overview of Evolutionary Mechanisms

Population genetics examines how the genetic makeup of populations changes over time, seeking to understand the mechanisms that drive adaptation and speciation. Four primary forces are recognized as drivers of evolutionary change:

- Genetic Drift
- Gene Flow
- Mutation
- Natural Selection

While all four mechanisms contribute to evolution, mutations and natural selection are by far the most frequently cited drivers of evolutionary change. In the sections that follow, we'll explore each mechanism and ask: Do they truly account for the complexity and diversity of life as we know it?

## Genetic Drift: Random or Meaningful?

Genetic drift refers to random changes in allele frequencies within a population due to chance events. This means that, purely by accident, certain versions of genes become more common, while others become less common. Unlike natural selection, which favors traits that enhance survival or reproduction, genetic drift is unpredictable. It is especially influential in small populations. Over time, it can reduce genetic diversity and cause certain traits to become fixed.

A classic example is the *founder effect*: when a small group splits off from a larger population, the gene pool of the new group is limited to the alleles carried by its founders. For instance, imagine a few birds blown off course to a remote island. These birds bring only a portion of the genetic variation from the mainland. Over generations, their isolated descendants may develop distinct traits simply due to random genetic sampling—not because those traits offer any survival advantage.

These changes illustrate genetic drift, not natural selection. The island birds may look or behave differently, but they remain the same species as their mainland counterparts. Moreover, because genetic drift operates independently of fitness, it lacks the directional force needed to consistently build new biological functions or complex structures. Thus, while genetic drift explains variation within species, it does not demonstrate the formation of entirely new species through macroevolution.

## Gene Flow: Migration, Not Macroevolution

Gene flow, or gene migration, occurs when individuals move between populations, introducing new genetic material. For example, immigration has increased genetic diversity in places like North America. While gene flow alters the genetic makeup of populations and promotes variation, it does not create new species. People from different ethnic backgrounds may have diverse traits, but they all share the same human genome—highlighting variation within a species, not the emergence of a new one.

## Mutations: Evolution's Catalyst or Hindrance?

Imagine every living organism as an intricately written genetic book, its pages filled with precise, life-sustaining instructions. Mutations are like sudden, random typos scattered throughout this book—caused by radiation, chemicals, viruses, or simple errors during DNA replication. These changes are often hailed as the raw material of evolution. But can random "typos" truly account for the emergence of entirely new traits or organisms?

Proponents of evolutionary theory argue that mutations, given enough time, drive genetic innovation. Yet in real biological systems, the story is often one not of progress, but of decay. Does error combined with blindness equal genius? *It does not.* That's not a scientific explanation—it's wishful thinking.

The notion that such genetic mishaps could fuel large-scale evolutionary leaps, from "amoeba to man," feels more like a leap of faith than a conclusion grounded in evidence. How can random errors be expected to craft complex, upward evolution?

Think of mutations like earthquakes: they may occasionally shift the ground harmlessly, but more often, they cause devastating destruction. Earthquakes don't improve buildings; likewise, mutations don't improve organisms.

Mutations generally fall into one of three categories: neutral, harmful, or—very rarely—beneficial.

## Near-Neutral Mutations

These are like shuffling a few letters in a sentence. The text may look slightly different, even confusing at times, but the overall meaning remains unchanged. Biologically, these are small DNA changes that have little or no effect on the organism. They're common—but don't drive evolutionary advancement.

## Harmful Mutations

Far more often, mutations are harmful. Picture a book caught in a fire—pages scorched, letters smeared, entire passages rendered unreadable. This is what happens when the genetic code is damaged: the organism may suffer from debilitating conditions, loss of function, or reduced ability to survive.

Radiation and toxic chemicals can scramble the DNA "letters" in exactly this way, sometimes leading to irreversible disorders like cancer. Just as a fire leaves permanent damage, so too do these agents leave a lasting, harmful imprint on the genetic code.

The rising prevalence of cancer and other genetic disorders suggests that harmful mutations are accumulating faster than natural selection can eliminate them. This trend poses a serious challenge to the idea of evolution as a consistently progressive force. Instead of driving improvement, this trend points to a gradual deterioration of genetic health over time.

## Beneficial Mutations

Mutations that are beneficial are exceedingly rare—and even then, they often come at a cost. They typically offer only minor, short-term advantages, not the kind of sweeping innovations required to explain complex biological systems.

A frequently cited example is **sickle-cell anemia**, often presented as a textbook case of a useful mutation because it offers resistance to malaria. Individuals with one copy of the mutated gene have some protection from malaria, as the parasite struggles to infect their misshapen red blood cells. On the surface, this appears to be an evolutionary advantage—but that benefit comes at a high price.

If a child inherits two copies of the sickle-cell gene, they develop full-blown sickle-cell disease—a serious, painful, and often life-shortening condition. They experience intense pain, organ damage, chronic fatigue, and an increased risk of stroke or early death. In malaria-prone populations, where the sickle-cell gene is common, roughly 25% of children born to two carriers will inherit both copies and suffer the full effects of the disease, while another 25% will inherit neither copy and thus gain no malaria resistance at all.

As Henry Morris and Gary Parker observed, even evolutionists acknowledge that such short-term advantages can result in long-term biological costs with detrimental consequences.[1] In other words, while the mutation may offer situational benefits, it introduces serious biological liabilities that challenge the argument for progressive evolutionary improvement through mutation alone.

If sickle-cell anemia were truly beneficial, populations in malaria-prone regions would embrace it. But no one seeks out this condition, because its long-term consequences are painful, dangerous, and damaging to health.

Ultimately, mutations remain unpredictable, mostly harmful, and fail to provide meaningful long-term benefit. Even the strongest example of a so-called beneficial mutation—sickle-cell anemia—does not convincingly demonstrate the creative power needed to drive large-scale evolutionary change. More often than not, mutations lead to dysfunction, decay, or death—not innovation.

## Can Natural Selection Come to Evolution's Rescue?

As we've seen, mutations are overwhelmingly harmful or neutral. But what about natural selection—can it swoop in, filter out the bad, and amplify the good to drive evolution forward?

## Natural Selection: Only Selects, It Doesn't Create

Natural selection—often summed up as "survival of the fittest"—is a process that preserves traits that enhance an organism's chances of survival and reproduction. But while it's commonly credited with driving evolution, natural selection doesn't create anything new. It merely selects from existing genetic variation.

These variations, introduced by mutations, may be neutral, harmful, or—in rare cases—beneficial. Since so-called beneficial mutations are extremely rare, natural selection is limited to sorting what already exists. It acts as a passive filter—not an active innovator. It can help organisms survive by preserving useful traits, but it cannot generate new organs, structures, or species.

At best, natural selection may slow genetic decline by favoring functionality. But it cannot reverse the overall trend of genetic deterioration caused by the ongoing accumulation of harmful mutations. For natural selection to power macroevolution, it would require a consistent supply of new, beneficial mutations—something we simply do not observe.

Macroevolutionary theory hinges on a mechanism that can reliably produce entirely new and functional genetic information.

Yet mutations are overwhelmingly random, typically harmful, and almost never truly beneficial. As such, they fall short of supplying the raw material required for large-scale evolutionary innovation.

The idea that blind, undirected genetic accidents could generate the intricate complexity found in every living organism runs counter to both scientific observation and reason.

If mutations truly served as the engine for upward biological progress—increasing complexity and genetic health over time—we would expect to see clear evidence of such a trend.

Instead, we observe the opposite: rising extinction rates, a growing list of genetic disorders, and a noticeable absence of the emergence of genuinely new species or higher taxonomic groups—particularly at the level macroevolution requires.

*If natural selection can only conserve, not create, what about artificial selection?* Could human-guided breeding succeed where nature falls short—and jumpstart the kind of transformation macroevolution requires?

## Artificial Selection: Can it Do What Nature Can't?

Artificial selection—including both traditional selective breeding and modern genetic engineering—allows humans to manipulate genetic traits with remarkable precision.

Yet even with these advanced techniques, we have never observed the creation of entirely new, complex, and functional genetic information—the kind required for macroevolutionary leaps—or the emergence of a new species.

This reveals a crucial limitation: neither artificial nor natural selection creates new traits. They can only work with the genetic material that already exists.

As author Vance Ferrell explains:

> *Selective breeding narrows the genetic pool; although it may have produced a nicer-appearing rose, at the same time it weakened the plant that grew that rose. Selective breeding may improve a selected trait, but it tends to weaken the whole organism. Because of this weakening factor, national and international organizations are now collecting and storing 'seed*

*banks' of primitive seed. It is feared that diseases may eventually wipe out our specialized [genetically altered] crops...*[2]

**This weakening effect highlights a major flaw: artificial selection, like natural selection, cannot cross the genetic boundaries that define a species.**

Breeders have produced dramatic variation in dogs—from tiny Chihuahuas to massive Great Danes—but these changes remain within the canine kind. No amount of selective breeding will ever produce a dog with a neck like a giraffe, or feathers like a bird—because the necessary genetic instructions simply aren't there. The genetic code imposes hard biological limits.

Despite its ability to enhance existing traits, artificial selection consistently hits a ceiling. It cannot invent new genetic blueprints.

To date, it has never produced a new species, much less a fundamentally new organism. Far from demonstrating macroevolution, artificial selection exposes its boundaries.

*If even guided, intelligent intervention cannot produce new kinds of organisms, what do real-world experiments reveal about evolution in action?*

## Fruit Flies: A Failed Test of Evolution

Mutations are simply too rare—and too often harmful—to account for the vast complexity and diversity of life.

For macroevolution to occur, many beneficial mutations would need to accumulate gradually over time. But decades of laboratory research tell a very different story.

One of the most famous test cases involves the fruit fly. Because fruit flies reproduce quickly and have just four chromosomes, scientists hoped they might showcase evolution in fast-forward. Throughout the 20th century, researchers bombarded them with radiation and chemicals, trying to trigger useful mutations.

As Richard Dawkins noted:

> *In laboratories worldwide, they have been subjected to all forms of mutation-inducing phenomena, including toxic chemicals and radiation treatments, in an attempt to accelerate evolution-mimicking mutations. After all this, fruit flies should have certainly exemplified evolution by now.*[3]

But they haven't.

Despite decades of effort, fruit flies have failed to evolve new structures or functional body parts.

Instead, researchers have produced deformities—flies with extra wings, misplaced legs, or disorganized body segments. These aren't signs of evolutionary progress; they're signs of biological breakdown.

The famous four-winged fruit fly is a prime example. Though visually striking, the extra wings lack flight muscles and are completely useless—making the fly less fit for survival, not more.

After countless generations of experimentation, fruit flies have never become anything other than fruit flies—just damaged versions of the original.

If evolution through mutation can't produce biological innovation under controlled lab conditions, what evidence is there that blind, undirected mutations can build the incredible complexity of life across our planet?

The same limitations become even more obvious in the microbial world—especially with bacteria.

## Bacterial Resistance and the Limits of Evolution

Bacterial evolution is often cited as clear evidence for macroevolution—particularly when bacteria develop resistance to antibiotics. But this is simply a textbook case of microevolution: small-scale variation within a species, not the emergence of a new one.

Consider the long-term experiment led by Richard Lenski at Michigan State University. For over 20 years and 40,000 generations, scientists observed changes in *E. coli* populations under controlled conditions. With bacteria reproducing so quickly, the experiment was expected to showcase evolution in high gear—**like evolution on steroids.**

But while mutations did occur, they were largely degenerative, resulting in functional losses rather than evolutionary gains. Despite tens of thousands of generations, no new species emerged.

If simple bacteria—reproducing rapidly under ideal laboratory conditions—fail to evolve into something fundamentally new, how can we expect far more complex organisms to do so in the wild?

This problem deepens when we examine *gene duplication*, often proposed as a pathway for new genetic information. But duplication merely copies what already exists.

As Dr. John C. Sanford, former Cornell geneticist and author of *Genetic Entropy*, explains, duplications frequently lead to genetic instability, not innovation.

Over time, harmful mutations accumulate, resulting in a net loss of information. This mirrors the second law of thermodynamics: systems tend toward disorder, not increasing complexity.

This leads to a critical question:

**If bacteria—the simplest, fastest-reproducing organisms on Earth—don't evolve into new species, what does that say about the plausibility of evolution from simple cells to complex life?**

Alan H. Linton, emeritus professor of bacteriology at the University of Bristol, put it this way:

> *But where is the experimental evidence? None exists in the literature claiming that one species has been shown to evolve into another. Bacteria, the simplest form of independent life, are ideal for this kind of study, with generation times of 20 to 30 minutes, and populations achieved after 18 hours. But throughout 150 years of the science of bacteriology, there is no evidence that one species of bacteria has changed into another, in spite of the fact that populations have been exposed to potent chemical and physical mutagens and that, uniquely, bacteria possess extra chromosomal, transmissible plasmids. Since there is no evidence for species changes between the simplest forms of unicellular life, it is not surprising that there is no evidence for evolution from prokaryotic to eukaryotic cells, let alone throughout the whole array of higher multicellular organisms.[4]*

Put simply: *If random mutations can supposedly transform one species into another, it's like claiming repeated car crashes could eventually turn a rusty junker into a high-performance sports car.* But common sense tells us: random crashes don't lead to engineering breakthroughs.

We see the same principle in biology.

Instead of genetic improvement over time, we're witnessing a rise in genetic disorders—a trend that suggests mutations are more destructive than constructive. As biophysicist Lee Spetner observes:

> *All point mutations studied at the molecular level reduce genetic information, not increase it.*[5]

If mutations truly built complexity, we would expect to see accumulating improvements.

On the contrary, both in humans and in bacteria, we observe genetic degradation. This starkly challenges the idea that random mutations, even when paired with natural selection, can account for the origin and complexity of life.

## The Biblical Perspective: Mutations, Entropy and the Curse

Mutations don't create—they corrupt. From a biblical standpoint, they are a consequence of the Curse described in

Genesis 3. Rather than building complexity, they erode it—leading to genetic defects, disease, and the gradual breakdown of the genome.

This pattern aligns not only with Scripture's portrayal of a fallen creation, but also with the scientific principle of **entropy**, the tendency of systems to move from order to disorder over time.

Just as physical systems wear down, so too does the genetic code—accumulating damage and dysfunction with each generation. Far from being engines of upward progress, mutations serve as markers of decline. Instead of driving evolution forward, they consistently reveal a loss of genetic information and biological integrity.

This echoes the biblical view: that creation began in a perfect state but has since been subjected to decay—a world now groaning under the weight of sin and entropy (Romans 8:20–22).

## Conclusion: The Missing Mechanism

Despite extensive study, no scientific mechanism has been conclusively demonstrated to generate the truly novel genetic information essential for one species to evolve into another. The widely cited processes of genetic drift, gene flow, mutation, and even natural and artificial selection may explain variation *within* species, but they utterly fail to account for the origin of new kinds of life. This profound and persistent gap decisively undermines the standard evolutionary explanation for life's vast diversity. The undeniable fact that evolution lacks a viable mechanism to drive macroevolution should serve as the final nail in its coffin.

This conclusion perfectly aligns with the biblical view that species were created with distinct genetic boundaries. Furthermore, it supports the understanding that mutations—introduced after the Fall—have resulted in a consistent loss of genetic information, rather than evolutionary gain.

If evolution's mechanisms cannot account for life's stunning complexity and inherent genetic blueprint, then what divine Architect brought forth existence and sustains its intricate design? This ultimate question, far beyond the realm of mere scientific mechanism, shapes our understanding of purpose, meaning, and reality itself.

But if evolution can't explain how new species arise over time, an even greater question remains:

***How did life begin in the first place?***

This leads us to the next challenge facing naturalistic theories—*the origin of life itself.*

# Can Life Arise from Non-Life?

*A scientist claimed, "The odds are just right,*
*That life came from nothing, with no guiding light."*
*I said, "If that's true, then here's my next task—*
*I'll try to build Rome with a rake and a flask!"*

If God created the world, then science is the study of how the world operates. Science helps us explore and understand creation—but it cannot tell us *why* there's a creation to explore in the first place.

That question marks the boundary of science—and reveals the greatest mystery we face: *How did life itself begin?*

## A Cell: More Complex Than a Jumbo Jet

A Boeing 747 Jumbo Jet consists of about six million parts. Each part is carefully designed, tested, and assembled with precision to ensure the aircraft functions as intended. Its complexity is a testament to human ingenuity.

By comparison, a single living cell—sometimes dismissed as "simple"—contains billions of molecular components, far more intricate and advanced than anything we've ever engineered. For perspective, the number of molecular parts in just one cell vastly exceeds the mechanical components of a jumbo jet.

A single human cell is composed of countless molecular parts, each with a highly specialized function. These components—proteins, organelles, and DNA—are not just random pieces but part of an intricately organized system working in perfect harmony to sustain life.[1]

***Complexity is never the product of confusion; it's the result of coordination.***

No one would reasonably argue that a 747 could have constructed itself solely by random processes. Such an idea seems absurd because we understand the necessity of intelligent design in creating complex machinery.

If a machine designed and built by humans can't assemble itself by chance, how could something as astronomically complex as a living cell do so on its own?

Yet, proponents of evolution claim that all the intricate components needed for life have somehow come together through natural processes—without any intelligent input or designer.

As discussed in previous chapters, while mutations and natural selection influence certain aspects of evolution, they fall far short of accounting for the astonishing complexity within even a single cell.

Without a guiding mechanism or intelligent input, these processes struggle to explain how such highly specialized, interdependent systems could have arisen by chance.

## The Odds of Life Emerging from Lifeless Matter

What is the likelihood of life originating from non-life?

Sir Fred Hoyle, the distinguished English mathematician and professor of astronomy, famously concluded:

> *The chance that higher life forms might have emerged in this way is comparable with the chance that a tornado sweeping through a junkyard might assemble a Boeing 747 from the materials therein.*[2]

What's even more striking is that Hoyle was not a creationist—he wasn't arguing from religious belief, but from sheer scientific skepticism. His objection wasn't theological; it was mathematical.

In other words, the odds of life emerging from non-living matter are even lower than the chances of the Minnesota Vikings winning their first Super Bowl—and I say that as a lifelong fan which means I know a thing or two about impossible odds.

But this isn't just a matter of probability—it's about process. At its core, life is not just chemistry—it's information. How could chemistry, by itself, give rise to code?

## From Dust to DNA? The Impossibility of Life Without Intelligence

Scientists often describe life as having emerged from the simplest elements—carbon, hydrogen, oxygen, and nitrogen—basic "dust" scattered across the universe. Yet the leap from these inert atoms to the stunning complexity of DNA, with its structured information and precise organization, defies natural explanation.

Random elements colliding is one thing; spontaneously producing the sophisticated molecules and coded instructions required for life is another entirely. Without an intelligent guiding force, the odds of life assembling from mere dust are astronomically low—so low, they challenge even the most generous estimates in mainstream science.

This isn't just a question of chemistry; it's a profound puzzle that points unmistakably to design and intelligence.

But how can we grasp the absurdity of the alternative? Perhaps a vivid illustration will help…

## The Blindfolded Painter

In a kingdom where art was everything, the king announced a contest:

"Whoever can recreate the Mona Lisa while blindfolded—using a paintball gun—will win half my kingdom."

Thousands lined up. The rules were simple: one blindfold, one canvas, one paintball gun loaded with random colors. Each contestant was spun in a circle and told to fire away.

Splats hit walls. Paint coated faces. One fellow accidentally shot a duck. The results were colorful chaos—but not a single Mona Lisa. Not even a stick figure. Not even a smiling blob.

Then a man stepped forward—eyes wide open, no blindfold. He studied da Vinci's masterpiece, mixed the paints by hand, and in hours recreated it with stunning precision.

The crowd gasped. "He cheated!" someone shouted. "He used intelligence!"

"Exactly," said the king. "Because even fools know—masterpieces don't happen by accident. They require a mind."

He handed the man the crown. "This is the only way art is ever made."

Much like the blindfolded painter's futile attempts to recreate a masterpiece, life—complex, purposeful, and finely tuned—cannot arise by accident. Random processes may produce chaos, but they cannot produce order. And certainly not the kind of intelligent, structured complexity we see in living systems.

The complexity of life, like the Mona Lisa, demands intelligence and design.

Can anyone seriously believe that chaos alone could produce such a masterpiece—not once, but billions of times over, in every living cell?

## Scientific Voices Against Random Evolution

Nobel Prize winner Dr. Francis Crick, co-discoverer of the molecular structure of DNA—one of the 20th century's most important biological discoveries—expressed serious doubts that life could have originated by chance on Earth. In his book *Life Itself*, Crick acknowledged that the origin of life appears to be "almost a miracle," given the precise conditions required for it to begin.[3]

Dr. Crick is not alone. Many respected scientists have likewise argued that life could not have arisen by mere chance, even when considering something as seemingly simple as a single functional protein, let alone an entire living cell.

A bacterium contains thousands of distinct proteins, each with a precise structure and function. If the random formation of even one protein is considered mathematically implausible by leading scientists, how likely is it that all the proteins required for life could have spontaneously assembled?

Zoologist Harold Coffin once emphasized the scale of this improbability, stating:

*The chances for producing the necessary molecules, amino acids, proteins, et cetera, for a cell (i.e. Mycoplasma hominis) is less than one in 10 to the power of 340,000,000 (or a 10 with 340 million zeros after it).[4]*

In mathematics, such a probability is effectively zero. According to **Borel's Law**, any event with a probability lower than 1 in 10 to the power of 50 is considered mathematically impossible.

Echoing this improbability, astrophysicist Sir Fred Hoyle remarked:

*The notion that the operating program of a living cell could be arrived at by chance in a primordial soup here on Earth is evidently nonsense of a high order.[5]*

This raises an even deeper question: not just how the parts of life could come together by chance, but how the instructions for life—its encoded information—could ever have written themselves.

## The Mystery of Biological Information

Dr. Wilder-Smith, a renowned chemist with three doctoral degrees, explained:

*An attempt to explain the formation of the genetic code from the chemical components of DNA is comparable to the assumption that the text of a book originates from the paper molecules on which the sentences appear, and not from any external source of information.*[6]

Dr. Wilder-Smith further emphasized:

*It is emphatically the case that life could NOT arise spontaneously in a primeval soup of this kind.*[7]

Former evolutionist Michael Denton, author and biochemist, also concluded that life cannot arise by chance:

*The complexity of the simplest known type of cell is so great that it is impossible to accept that such an object could have been thrown together suddenly by some kind of freakish, vastly improbable event. Such an occurrence would be indistinguishable from a miracle.*[8]

All these voices—from Crick to Hoyle to Denton—converge on the same mystery. The real question is not merely how life's building blocks assembled, but how they came to carry meaning:

**Where did the information come from?**

**How did the instructions for life—its genetic code—ever write themselves?**

*You don't leap from chemicals to code by accident.*

## The Role of Natural Selection

Some critics argue that evolution is not purely a chance process and that natural selection improves the odds. Yet natural selection operates only within living organisms—not among inanimate chemicals, the very materials evolutionists claim preceded life.

As we discussed in earlier chapters, natural selection functions as a passive filter. It can preserve traits that offer an advantage, but it cannot create new genetic information or produce entirely new species.

Instead of viewing evolution as a process from "*amoeba* to man," we might take a broader view—from "particle to man." After all, life didn't begin with living cells, but with lifeless matter. The real challenge is not how species evolved, but how life began.

As author **James Perloff** humorously observed:

> *No one wins this race for survival—all the contestants are dead at the starting line.*[9]

This raises an even more fundamental question:

*If natural selection can only act on life that already exists, what mechanism—if any—can account for the origin of the very first living system?*

## From Cake Batter to Consciousness?

Let's assume, for the sake of argument, that dumb luck could somehow assemble amino acids into proteins. Would that alone create a living organism? Once again, James Perloff offers a fitting analogy:

> *We can drop sugar, water, flour, baking powder, and an egg on the floor, but they won't turn into a cake by*

*themselves. We have to mix and bake them according to a recipe.*[10]

In the same way, combining hydrogen, helium, and oxygen in outer space will never accidentally produce a living organism. Even though the universe is made of the same fundamental elements as stars, it takes a **divine genius** to arrange that stardust into something as intricate and astonishing as life.

## The Mosquito Analogy

Imagine tossing a thousand mosquitoes into a blender and liquefying them into a "protein shake." What are the chances those mosquito molecules could ever spontaneously reassemble into a swarm of flying bloodsuckers? Even if every chemical component of those mosquitoes remains in the blender, not one will ever re-form into a living creature.

All the ingredients are still there, yet no mosquito reemerges— because life is not merely about ingredients. It's about precise assembly and encoded instruction.

It is like scattering the parts of a broken Rolex into a box and expecting a working watch to appear just by shaking it long enough.

In fact, nothing beautiful or sophisticated will ever arise from a blender full of mosquito soup—no matter how long we wait.[11]

Tom Wagner, a science teacher and author, aptly pointed out that *"A squashed mosquito is dead forever."* He skillfully illustrates the point:

> *Interestingly, the squashing of a mosquito may help us understand what makes life possible and what makes the spontaneous generation of life impossible.*
>
> *When a mosquito is slapped, what happens? Obviously, its shape changes and it dies. But what makes it die? All of the thousands of sophisticated chemicals which make up its body are still there, relatively unaltered. At the moment of impact, its cellular components are still intact including the all-important DNA. So why is it now dead?*
>
> *Prior to being smashed, the mosquito was a highly organized system with complex information. But when it got hit, it became disordered, causing critical information in the design of its body to become jumbled. There arose confusion in the finely tuned coordination of chemistry (including the chemicals involved in its overall structure) which culminated in an overall breakdown, resulting in death.[12]*

**If the brightest scientists in the world have failed to create life on purpose, how could blind chance accomplish it by**

**accident?** Let's be honest—unthinking chemicals cannot turn themselves into thinking human beings.

This raises an important question: if engineers receive credit for machines, and software developers are praised for writing code, why are we reluctant to credit a Designer for something exponentially more intricate—like DNA? We recognize intelligence when we see it. We admire the minds behind skyscrapers, smartphones, and satellites.

*So why do we stop short when we encounter the greatest engineering feat of all: life itself?*

## The Designer Behind Life

Dr. Michael Behe, Professor of Biological Sciences at Lehigh University, made a compelling case against Darwin's theory in his book *The Edge of Evolution*. In his conclusion, which is prominently displayed on the back cover of the book, he states simply:

### *The universe was designed for life.* [13]

A universe designed for life stands in direct contrast to the idea of random chance evolution. Speaking of random chance, here's a true story about an evolutionist who miscalculated and misled his audience. Thomas Huxley, known as **"Darwin's Bulldog,"** falsely claimed that six monkeys randomly typing on typewriters for a million years could write all the books in the British Museum. James Perloff debunked this claim:

> *Anyone who believes these projections hasn't done the math. What are the odds of a monkey typing a predetermined nine-letter word like 'evolution'? With 26 letters in the alphabet, the monkey would need more than five trillion attempts to type 'evolution' once correctly. At ten letters per minute, this would take over a million years. If typing one word is that difficult, imagine the challenge of producing a paragraph...[14]*

If random typing can't form a paragraph, it certainly can't build a genome—the most sophisticated information system ever discovered.

If design implies intelligence, then nowhere is that intelligence more evident than in the very language of life itself—DNA.

## The Information in DNA

Now, imagine the probability of producing and organizing all the quadrillions of components that define life. **It's staggering.** Saying life formed by accident is like expecting an explosion of Scrabble tiles to spell out the *Declaration of Independence*—verbatim, with perfect punctuation.

As George Sim Johnson wrote in the *Wall Street Journal*:

*Human DNA contains more organized information than the Encyclopedia Britannica. If the full text of the encyclopedia arrived in computer code from outer space, most people would regard it as proof of extraterrestrial intelligence. Yet, when this information is found in nature, it's explained as the result of random chance.*[15]

DNA is the most densely packed assembly of information in the known universe. It must be the work of a master designer—yet secular evolutionists refuse to acknowledge this.

The gene segments that connect all DNA helix strands are represented by a string of four chemical letters. These letters—**C, G, A,** and **T**—are arranged in a precise, coded sequence to produce the essential instructions that make life possible.

To put this in perspective, consider your smartphone. Every app, every picture, every function it performs is the result of carefully written code. **Now imagine compressing that same level of complexity—millions of lines of code—into something as tiny as a single cell. That's what we see in DNA.**

Just as your phone's operating system couldn't come into existence by accident, neither could the highly organized information in DNA.

It takes an intelligent designer to write genetic code—just as it takes an engineer to write software code.

If chance were the author of life, how could it hit the bullseye—*every single time, in every cell, across every species?*

## The Complexity of Life's Blueprint

What does it take to build a functional automobile? It requires an intelligent, creative mind. Who gets credit for the creation of a car—the steel frame, the leather seats, the rubber tires, the glass windows, the gasoline? Of course not. It's the designers, engineers, and builders who deserve credit.

So, when it comes to complex, sophisticated living organisms, why do evolutionists credit the parts instead of the Creator?

Dr. Walt Brown, in *In the Beginning*, offers an amazing description of DNA:

> *If all the DNA in one of your cells were uncoiled,*
> *connected, and stretched out, it would be about 7 feet*
> *long. If all this densely coded information from one cell*
> *were written in books, it would fill a library of about*
> *4,000 books. If all the DNA in your body were stretched*
> *end-to-end, it would reach the moon more than 500,000*
> *times! In book form, this information would fill the*
> *Grand Canyon almost 100 times. If you gathered all the*

*DNA from every person who has ever lived, the pile*
*would weigh less than an aspirin!*[16]

All this information is stored and functioning inside a microscopic cell, each one working flawlessly without a single error. Think of it like a master architect designing the blueprint for a skyscraper—every floor, every support beam, every wire and pipe has a precise place and function. Now imagine that design packed into a space smaller than a sugar cube, yet infinitely more complex. **What a magnificent Creator** to craft such intricate order!

That's DNA: an astonishingly detailed blueprint for life, written in a language far more intricate and sophisticated than any human-made computer code.

Understanding DNA is just one small reason for believing that you are *"fearfully and wonderfully made"* (Psalm 139:14).

## The Myth of Junk DNA

For those critics who once argued that our cells contain "junk DNA," *The Myth of Junk DNA* by Jonathan Wells is a valuable resource that thoroughly debunks this misconception.

Despite the overwhelming evidence of high-quality, information-rich DNA, many with a strictly materialistic worldview insist that everything—including life—can be explained by physics and chemistry alone. But Dr. Werner Gitt, an information scientist, disagrees:

*It should be noted that the activities of all living organisms are controlled by programs comprising information. Because information is required for all life processes, it can be stated unequivocally that information is an essential characteristic of all life. All efforts to explain life processes solely in terms of physics and chemistry will always be unsuccessful. This is the fundamental problem confronting present-day biology, which is based on evolution.[17]*

Bacteria lack the genetic information to produce eyes, ears, hands, and feet. Humans, however, possess this vast amount of information. The question is: where did it come from? Evolutionists have no answer. There is no known mechanism by which meaningful, complex information can randomly appear. Since every living organism is packed with meaningful genetic information, there is absolutely no way the creatures populating the earth today could have come about by pure chance.

In the end, energy and matter are material—but information is not. Dr. Gitt continues:

*There is no known law of nature, no known process, and no sequence of events which can cause information to originate by itself in matter.[18]*

## The Mystery of the Genome

Geneticist John Sanford remarked:

*As we recognize the higher-ordered dimensions of the genome, we can agree with Carl Sagan (1974) that each cell contains more information than the Library of Congress. Human life is more complex than all human technologies combined!* ***Where did all this information come from, and how can it be maintained? This is the mystery of the genome.[19]***

The sheer complexity and volume of information encoded within the genome challenge the notion that it could arise from mere chance or random processes. The genome is not just a collection of biological material; it is a finely tuned, highly sophisticated system that stores, processes, and transmits the very blueprint of life.

The precision with which it operates suggests a level of design far beyond what random chance could achieve.

This kind of complexity points to more than just chance—it strongly hints at a purposeful origin. The intricate, life-sustaining systems within the genome reflect the work of an intelligence capable of crafting such remarkable mechanisms.

And as our understanding of genetics deepens, the evidence grows clearer: *the genome is not the product of blind, undirected forces, but of careful and intentional design.*

## A Personal Invitation

The more we understand about life's origins, the more we see evidence not of blind chance, but of purposeful design. And if life is the product of intention, perhaps we too are meant for something far greater than mere accidents. The intricacy of the genetic code, the order within cellular systems, and the sheer volume of encoded information all point toward a Designer who works with meaning and careful purpose.

So where does all this lead? If DNA is more than chemistry—if life is more than molecules—then maybe we are more than accidents. Perhaps there is a Designer who not only engineered life with breathtaking complexity but also formed each of us with deliberate care.

Could it be that you're here on purpose? That your life—like DNA—isn't an accident, but a message crafted just for you, waiting to be discovered?

If you've begun to sense that life's origin points to something greater, don't stop here. In the final chapter, Stumbling Block #16: *The Bible vs. Evolution: The Ultimate Collision*, we invite you to discover the identity of that Designer—not just as Creator, but as Savior. The truth about our origin leads to the truth about our purpose. And when you meet the One behind it all, it just might change everything.

*Turn the page. Discover the truth science can't explain—but your soul was made to find.*

# Five Scientific Laws

# Evolution Can't Escape

*They say evolution is a scientific feat—*

*That life came from nothing, a trick quite neat.*

*But the laws of physics, precise and grand,*

*Say, "Something from nothing? That won't stand!"*

## When Scientific Laws and Evolution Collide

Evolution is widely accepted as a scientific theory—but why does it seem to clash with some of science's most established laws?

Science rests on observations, consistency and repeatable testing. Yet when we set evolution beside several of the universe's foundational laws, the tensions become hard to ignore.

In fact, five fundamental laws of nature appear to challenge evolution's core claims directly.

To understand why, we need to distinguish between *laws* and *theories*. Laws describe patterns in nature that have been repeatedly confirmed, while theories attempt to explain *why* those patterns occur. The differences matter.

Theories are like someone who talks a big game but never quite delivers—speculative, ever-changing, and still under review. Laws, on the other hand, are like your grandma—reliable, proven, and you don't argue with them.

She may not change with the trends, but that's exactly what makes her trustworthy. In science, laws don't yield to theories.

So why do we sometimes bend the rules for evolution—a theory that struggles to stand up under its own weight?

This isn't just a technical debate—it's a clash of worldviews. What's at stake is not only how we interpret the scientific evidence, but also what we believe about our origins and our purpose.

Before exploring these challenges, we need to ask a deeper question:

Where do these universal laws—the very structure of reality—come from?

Are they the accidental byproduct of a chaotic universe, or does their order and precision point to the work of a greater Lawgiver?

This question sets the stage for why these laws pose such serious trials to evolutionary theory.

## Five Laws That Put Evolution to the Test

Let's examine five scientific laws that raise critical questions about the theory of evolution. Don't worry about the technical names—we'll explain each one clearly and simply:

- Conservation of Energy (1st Law of Thermodynamics)
- Entropy (2nd Law of Thermodynamics)
- The Law of Biogenesis
- Borel's Law
- The Law of Cause and Effect

# The First Law of Thermodynamics: Conservation of Energy

The First Law of Thermodynamics states that energy cannot be created or destroyed—only transformed from one form to another.[1]

Since matter and energy are interchangeable (as Einstein's famous equation, $E = mc^2$, shows), this law applies to both.

In every process scientists have observed, energy and matter never appear from nowhere—they're always rearranged or converted.

Even with the most advanced technology, humans can't create a single atom or grain of sand from absolute nothing. We can only work with what already exists.

But this leads to a deeper question: if matter and energy cannot create themselves, where did everything come from?

The First Law describes what happens inside the universe—it assumes the system already exists. But it doesn't explain how the universe itself began. That's a major limitation.

If physical laws can't account for the origin of the very system they govern, the discussion naturally shifts beyond science, into philosophy or theology.

Some suggest the universe created itself, or sprang into existence from a quantum fluctuation or an undefined "nothing."

But these ideas don't hold up under scrutiny.

The First Law doesn't permit energy or matter to arise from nothing. It clearly tells us that "something from nothing" is not how the universe works.

So when we hear that the universe emerged spontaneously—no energy, no matter, no Creator—just a sudden "poof" from the void, we have to ask: where did it all come from?

Unless "nothing" can create everything, a better explanation is needed.

The most reasonable conclusion—consistent with both science and logic—is that the universe had an intelligent, powerful cause: a divine Creator.

The First Law reminds us that matter and energy don't just appear without cause—and the universe is no exception.

And this is only the beginning. Another well-established scientific law poses an equally serious challenge to evolution.

## The Second Law of Thermodynamics: Entropy

The Second Law of Thermodynamics deals with entropy—the natural tendency of all systems to move from order to disorder.

Imagine an old book left on a shelf for decades. Its pages yellow, ink fades, and the paper crumbles into dust. That's entropy in action—everything breaks down over time, including genes, cars, and even our bodies.

Entropy is the universe's default setting: things naturally move toward disorder, not order.

Whether it's rusting metal, aging bodies, or crumbling buildings, time flows one way—toward decay, not spontaneous improvement.

When you think about entropy, remember this simple truth: **"Decay is the Way."**

But evolution presents a serious problem. It suggests simple molecules became more complex and organized, apparently defying entropy's law of disorder.

Evolution claims that order and complexity spontaneously arise from chaos—but entropy is a one-way street.

Leave a Cadillac in a field for a thousand years, and it will slowly disintegrate into rust, dirt, and broken parts. That pile of rusty debris will never "evolve" back into a polished Cadillac.

This principle applies universally: random collections of parts—like scrap metal or debris—don't self-organize into complex, functioning systems.

Yet evolution claims that complex life forms emerged from disordered molecules over millions of years. That's not just improbable; it contradicts everything we know about the universe.

Supporters of evolution often respond that Earth isn't a closed system. They argue that external energy sources—like the sun—allow local decreases in entropy, such as trees growing or babies developing, without violating the Second Law, since total entropy still increases overall.

That's true—but there's a deeper issue. Local decreases in entropy don't occur just because energy is present. They require pre-organized systems, coded instructions, and mechanisms—like DNA, molecular machines, and cellular frameworks.

Without these, energy alone doesn't create complexity; it creates chaos.

You could shine sunlight on a pile of bricks for a billion years, and it still wouldn't build a cathedral.

Likewise, raw energy pouring into non-living matter doesn't explain how the first self-replicating, information-bearing systems came into existence.

Even with Earth's access to external energy, the origin of life remains a mystery that random chance and natural selection cannot explain.

The Second Law reminds us that the universe naturally trends toward disorder, not toward functional, information-rich order. Without a guiding force, life doesn't just "happen"—not chemically, not biologically, not statistically.

From this perspective, the Bible offers insight into the origin of order and the introduction of decay.

According to God's Word, there was no entropy in the Garden of Eden—everything was in perfect order.

But the moment Adam and Eve disobeyed and ate from the tree of the knowledge of good and evil, the entire creation came under entropy's curse.

Since then, disorder has steadily increased—and will continue until Jesus returns.

The Bible speaks directly to this decay in Romans 8:20–21:

*For the creation was subjected to frustration, not by its own choice, but by the will of the one who subjected it, in hope that the creation itself will be liberated from its bondage to decay...*

Psalm 102:26 echoes this truth, explaining that all people will wear out like a garment. And in Romans 8:22, the apostle Paul writes:

*We know that the whole creation has been groaning as in the pains of childbirth right up to the present time.*

Ultimately, the Second Law remains fully intact, underscoring a simple reality: disorder—not order—is the natural direction of all systems.

The Second Law of Thermodynamics doesn't care about Darwin's theory.

**Entropy always wins.**

That's not a theological opinion—it's a scientific reality that no evolutionary storytelling can override.

## The Law of Biogenesis

The Law of Biogenesis states that life comes only from pre-existing life. Simply put, life does not spontaneously emerge from non-living matter.

Despite decades of experimentation—including many attempts to recreate early Earth conditions in so-called "primordial soup" experiments—scientists have never observed life forming from inorganic chemicals under natural conditions.

The origin of life remains one of science's great unanswered questions.

But this principle stands clear: **once something is truly dead, it stays dead.**

Consider this: when a person dies, every atom in their body is still there—but life itself is gone.

No medical technology, no matter how advanced, can bring it back. Not even a metaphorical *"Darwinian defibrillator"* can restore life once it's lost.

Death is final.

The Bible records miraculous moments where life returns after true death—like Lazarus, or the resurrection of Jesus Christ. But these events are attributed solely to divine power, underscoring that life-from-death is not a natural process, but a supernatural act.

If life only comes from life, how can we believe it began from a random collision of chemicals, billions of years ago?

It may sound like the premise of a science fiction novel—but in the real world of observable science, there is no empirical evidence that life can come from non-life.

Until proven otherwise, the most reasonable conclusion—supported by observation and history—is that life comes only from life.

This isn't theoretical guesswork; it's an established scientific law.

Some researchers argue that life might have emerged under unknown conditions on early Earth. The idea of a "primordial soup" giving rise to life is often repeated, but it remains highly speculative, lacks experimental confirmation, and defies common sense. Despite its imaginative appeal, it has not been verified in any scientific setting.

The Law of Biogenesis—confirmed through centuries of careful observation—still holds.

Until credible, observable, and repeatable evidence contradicts it, the claim that life emerged from non-life remains what it's always been: speculation, not science.

As if the challenges of energy, order, and origin of life weren't enough, evolution also faces an astronomical hurdle when confronted with the laws of probability.

## Borel's Law

Borel's Law is a mathematical principle. It states that if the probability of an event is less than one in ten to the power of fifty—a 1 followed by fifty zeros—it's considered mathematically impossible.[2]

Now consider the odds of forming the molecules needed for a single living cell. The probability is estimated at one in ten to the power of 340 million—a 1 followed by 340 million zeros. It's nearly beyond comprehension—a number so massive, it stretches the limits of possibility.

The chances of life emerging by random chance are so astronomically small, they're effectively zero. You'd have better luck winning the lottery every single day for the next million years.

Even under the most optimistic assumptions, the odds of assembling even the simplest cell by random processes are so extreme that many scientists acknowledge the need for a guiding principle—or Designer—behind life's origin.

Yet evolution asks us to believe that such an improbable event occurred—not just once, but *countless times*—producing the stunning complexity of life we see today. That's not merely unlikely—it's mathematically impossible.

Some argue that time is the great equalizer, suggesting that billions of years could make the improbable inevitable. But time doesn't solve the problem—it magnifies it. Evolution wields "time" like a magic wand, yet entropy moves in the opposite direction. More time only brings more disorder, not complexity. Time isn't the hero of the story; it's the villain.

Others invoke the multiverse theory, proposing infinite universes where life might arise from non-living matter. Intriguing as it sounds, it merely "kicks the can down the road," offering speculation, not proof. This theoretical escape hatch shifts the problem to an unverifiable realm where anything becomes possible—and nothing is testable.

Ultimately, these ideas—whether invoking vast time or infinite universes—remain just that: speculation. They only obscure the glaring reality that the odds of life arising by chance are so small they border on the impossible.

And no amount of imagination can change that fundamental truth.

Next, we turn to what may be the most intuitive and undeniable law in all of science—one that speaks directly to the universe's origin and dismantles the idea of creation from nothing.

## The Law of Cause and Effect

The Law of Cause and Effect is one of the most universal principles in all of science—and in life. It simply states that every effect must have an adequate cause.[3]

From physics to chemistry to everyday experience, this law holds true. A painting requires a painter. A song requires a songwriter. A design requires a designer. **Cause and effect are inseparable.**

So what caused the universe?

If everything we observe has a cause, then logic demands the universe itself must have one too. Something—or someone—had to start it all.

But evolution and naturalism insist that the universe, and life within it, somehow caused themselves. In this view, **everything came from nothing**, without a cause, without intelligence, and without purpose.

Yet this claim breaks the very law that science depends on to function.

The universe cannot be both the **effect** and the **cause** of itself. That's like saying a book wrote its own story, or that a match lit itself before it existed. Such ideas defy logic, science, and reason.

Even leading atheists and cosmologists have acknowledged this dilemma. The laws of physics describe how the universe operates—but they don't explain how the universe began. The question of ultimate cause still stands unanswered.

Every law, every constant, every force in nature points back to an intelligent source that set them in motion.

If there was ever a time when **absolutely nothing** existed, then nothing could exist now—because nothing produces nothing.

Something—or rather, **Someone**—had to exist before everything else.

Someone powerful enough to cause all matter, energy, and life to come into being.

That Cause is not an accident, a random fluctuation, or a mathematical anomaly. The only explanation that aligns with both logic and observation is an eternal, intelligent Creator—the uncaused Cause behind all causes.

The Law of Cause and Effect doesn't just challenge evolution; it points beyond it—to the One who spoke the universe into existence.

## Evolution: A Leap of Faith?

These five scientific laws aren't fringe theories or theological add-ons—they're the bedrock of how we understand reality itself.

And when we examine them honestly, they don't point to chaos or chance—but to order, intention, and design.

Evolution asks us to accept a chain of impossible events, each without direction, purpose, or cause.

But creation begins with something far more logical—and far more personal:

A Cause worthy of the effect.

A Lawgiver behind the laws.

A Mind behind the code.

A Designer behind the design.

*The Creator is not a crutch for the uninformed—He is the only explanation that upholds the very laws science is built upon.*

So the question remains: Will we keep forcing the laws of nature to fit a theory that defies them—or follow the evidence to the One who wrote those laws in the first place?

# Can We Trust the Bible?

*They've poked and prodded every page,*

*From ancient scrolls to skeptic's stage.*

*While ages pass and doubters aim,*

*Its truth endures—still just the same.*

*From kings and scrolls in distant lands,*

*To words fulfilled by sovereign hands.*

*This book, though old, won't crack or crumble—*

*Its roots run deep, it doesn't stumble.*

Before we can meaningfully explore what the Bible says about origins, creation, and humanity's purpose, we must first settle a foundational question: ***Can we trust the Bible itself?***
If the Bible is unreliable, then its message about life, morality, and destiny carries little weight. But if it is historically and divinely trustworthy, then it provides the firm foundation on which every other truth rests.

## The Bible Under Fire—And Still Standing

Just as the poem suggests, despite scrutiny through the ages, the Bible's truth remains unshaken.

Let's explore why this ancient book continues to endure in the face of doubt.

## A Book Built on Evidence

To build a foundation for trusting the Bible, we must first examine the evidence behind its reliability.

The Bible is not merely an ancient collection of stories—it stands on a bedrock of *historical documentation, archaeological findings, manuscript integrity, and fulfilled prophecy*. These four pillars of evidence converge to demonstrate that the Bible is a credible and trustworthy source, worthy of serious consideration.

## A Legacy of Lives Transformed

Beyond the factual evidence, millions of lives throughout history and across the globe have been transformed by its message. This enduring impact serves as a powerful, living testimony to its authenticity.

## Facing the Questions Honestly

Understanding why the Bible deserves our confidence is essential, especially for skeptics who question its credibility.

This exploration will highlight how the Bible has withstood intense scrutiny, rigorous analysis, and more criticism than any other book in human history—yet remains steadfast.

## A Resource for the Curious

For a deeper investigation, *The Case for Christ* by Lee Strobel offers a well-researched and accessible examination of historical and manuscript evidence, presenting a compelling case for the Bible's authenticity.

It provides skeptics with a thoughtful foundation for trust.

## Where We Begin

With that foundation set, we now begin by examining the Bible's historical reliability.

## Historical Evidence

One of the Bible's most compelling strengths is its remarkable historical accuracy. It consistently names real people, places, events, and timelines with impressive precision—many of which have been verified by secular historians and independent sources.

Even critics who once dismissed the Bible as mere myth have had to acknowledge its surprising consistency with history.

Consider the writings of **Josephus,** a first-century Jewish historian who was not a Christian, yet often confirmed the existence of key biblical figures and events.[1] Other non-Christian sources also support the biblical record—Roman historians such as Tacitus and Pliny the Younger referenced Jesus and the early Christian movement in their writings. In addition, numerous inscriptions, official records, and archaeological discoveries outside the Bible continue to validate details that were once questioned or dismissed.

A powerful case in point is the biblical account of **King David**. For many years, skeptics regarded David as a mythical figure. That changed in the 1990s with the discovery of the Tel Dan Stele—a 9th-century BC inscription referring to the "House of David." This artifact provided concrete evidence that David was not a legend, but a real historical king, as described in the Bible.

The Bible's historical reliability is also reinforced by the eyewitness nature of many of its writings. The New Testament, in particular, was written by individuals who personally witnessed the life, death, and resurrection of Jesus.

In historical research, eyewitness testimony is one of the most trusted forms of evidence—lending substantial credibility to the biblical narrative.

Remarkably, even Jesus' enemies did not deny His miracles—they sought to explain His power as demonic (Matthew 12:24). This indirect admission reveals that something extraordinary was happening—something even His critics couldn't ignore. Their opposition actually affirms the reality of His works, indicating they weren't later inventions, but undeniable moments that demanded explanation.

Perhaps even more astonishing is the internal consistency of the Bible. Spanning over 1,500 years and written by more than 40 authors from a wide range of cultures, occupations, and locations, the Bible still presents a cohesive and unified message.

This level of harmony across such diverse contexts is virtually unmatched in ancient literature, and it further supports the Bible's authenticity as a reliable historical document.

## Archaeological Evidence

Archaeology has long been a powerful tool for understanding ancient civilizations—and it has played a significant role in affirming the credibility of the Bible.

While some early archaeologists set out to disprove Scripture, what they uncovered often challenged their assumptions. Decades of excavation and study have revealed discoveries that closely align with the biblical record.

To this day, no major archaeological discovery has conclusively disproven any biblical claim. On the contrary, new discoveries continue to affirm its historical accuracy.

A striking example is **Sir William Ramsay**, a respected archaeologist from the 19th century. At first, Ramsay was convinced that the New Testament—especially the writings of Luke—could not be trusted historically.

After years of careful research and fieldwork in the Middle East, he concluded that the details recorded in the book of Acts were remarkably accurate—too precise and consistent to ignore.

His journey from skepticism to belief underscores a recurring theme: honest investigation often leads toward faith, not away from it.[2]

Other archaeologists have experienced similar shifts in perspective. Discoveries in ancient **Jericho**, the unearthing of long-lost cities like **Nineveh** and **Capernaum**, and inscriptions confirming historical figures such as **Hezekiah**, **Caiaphas**, and **Pontius Pilate** have all added weight to the Bible's historical claims.

The connection between Scripture and real-world history continues to be confirmed through ongoing excavations. From the rediscovery of ancient cities to findings that align with key events like the **Exodus**, archaeology continues to reinforce the Bible's reliability.

These discoveries confirm that the Bible's narrative is grounded in real history—not myth or legend.

With every new finding, the message becomes clearer:

*The Bible doesn't exist in a vacuum of faith—it is anchored in the soil of history.*

## Manuscript Evidence

Among all ancient texts, the Bible stands as the most reliably verified. The New Testament alone has more than **24,000** manuscript copies—more than any other ancient document by a wide margin. Some of the earliest New Testament manuscripts date to within **25 to 45 years** of the original writings, providing an unusually short gap between the events and their documentation.

The Old Testament also has a strong manuscript tradition, with more than 10,000 known copies. This makes it the second most well-attested ancient text in terms of manuscript numbers, far surpassing other classical works.

For comparison, other respected ancient works—such as *Homer's Iliad*, which holds the third-highest number of surviving manuscripts—have only **643 copies**, and the earliest of those were written around **500 years** after the original.

Others, like *Plato's Dialogues*, are based on just **seven** known manuscripts, with a time gap of more than **1,200 years** between the originals and the earliest copies. Yet these texts are still considered historically reliable. By contrast, the New Testament's manuscript evidence far surpasses it—in both quantity and proximity to the original events.

A significant boost to the Old Testament's manuscript support came with the discovery of the **Dead Sea Scrolls** in the mid-20th century. Found in caves near Qumran—an archaeological site by the Dead Sea—these ancient manuscripts date from the 3rd century BC to the 1st century AD, making them over a thousand years older than the previously oldest complete Hebrew manuscripts.[3]

What they revealed was remarkable: despite the massive time gap, the Dead Sea Scrolls showed exceptional consistency with the Masoretic Text—the traditional Hebrew Bible used in modern translations.[4]

While minor variations in spelling and grammar exist, the core message and content remained virtually unchanged over more than a thousand years of transmission. This discovery provides compelling evidence of the Old Testament's careful and accurate preservation.[5]

Skeptics often point to copying errors in ancient texts. And while minor errors do appear in some biblical manuscripts, scholars are able to compare thousands of copies, identify discrepancies, and reconstruct the original text with remarkable accuracy.

This process is guided by the field of **textual criticism**—a rigorous, global, and multi-disciplinary effort involving scholars from diverse backgrounds and denominations (including skeptics and non-religious academics). These experts analyze and compare manuscript evidence to determine the most reliable wording of the original texts, using established historical and linguistic methods.

To illustrate, imagine ten people transcribing the same page from an encyclopedia. If nine copies are identical and one has a spelling mistake, the error is easy to spot. In the same way, with over 24,000 manuscripts of the New Testament, scholars can confidently distinguish genuine content from copyist errors.

In fact, well under one percent of the text remains uncertain—and most of these involve minor grammatical or spelling variations. None of them affect any core doctrines or theological teachings.

Even if every manuscript of the New Testament were lost, scholars could reconstruct the entire text using over one million quotations from early Christian writings and sermons.

This unparalleled manuscript tradition makes the Bible not only the most well-preserved document of the ancient world but also the most reliably transmitted.

## Prophetic Evidence

One of the most remarkable aspects of the Bible is its prophetic accuracy. Many Old Testament prophecies—written centuries before the events they describe—were fulfilled in the life and death of Jesus Christ.

These prophecies are specific, detailed, and fulfilled in ways that could not have been staged or self-fulfilled, strongly pointing to divine inspiration.

Take the book of Isaiah, written around 700 years before Jesus' birth. It contains vivid descriptions of the Messiah's suffering, death, and resurrection. These were recorded long before the events occurred, and the details are so precise that fabrication after the fact would be virtually impossible.

A common objection is that these prophecies were written after the events occurred.

However, this theory is significantly undermined by the discovery of the Dead Sea Scrolls. Found in the mid-20th century and dated to at least 100 years before Jesus' birth, these ancient manuscripts contain many of the same Messianic prophecies—including those in Isaiah 53. Their existence confirms that these prophetic texts were in circulation long before the events of the New Testament took place.

Other notable prophecies include the destruction and future rebuilding of the Jerusalem temple and the restoration of Israel as a nation—events that unfolded centuries after the original prophecies were given, and which continue to be significant to biblical prophecy today.

This archaeological evidence strongly supports the authenticity and divine origin of biblical prophecy.

Scholars have identified more than **300 distinct Old Testament prophecies** that were fulfilled in the life of Christ. The statistical odds of even a few of these prophecies coming true in one person purely by chance are astronomically low.

According to mathematician Peter Stoner, the odds of just eight of these prophecies being fulfilled in one person are one in ten to the 17th power—**that's a 1 followed by 17 zeros.**[6]

To help visualize this, Stoner offered a striking illustration: Imagine covering the entire state of Texas with silver dollars, two feet deep. Then mark one coin, mix them all up, and blindfold a person to pick a coin at random. The chance of them picking the marked coin on their first try is the same as just **eight** Messianic prophecies being fulfilled in one man by random chance.

The unity of these fulfilled prophecies—woven across centuries of Scripture—reveals a consistent divine voice, not disconnected predictions.

Taken together, these fulfillments form one of the most compelling lines of evidence for the Bible's divine origin.

Consider just a few remarkable prophecies: His birth in Bethlehem... His betrayal for thirty pieces of silver... the casting of lots for His clothing... the method of His execution... the piercing of His side... and His burial in a rich man's tomb.

All were foretold in Scripture and fulfilled in the New Testament accounts. These are not vague or symbolic predictions—they are clear, specific, and historically verified—prophecies that could not have been coincidental.

The sheer volume, detail, and accuracy of these fulfilled prophecies place the Bible in a category of its own.

No other religious text matches this level of prophetic fulfillment—not the Quran, the Bhagavad Gita, or any other sacred book—making the Bible uniquely trustworthy and unmistakably divinely inspired.

## The Bible's Radical Transparency

In addition to the many forms of evidence already explored, another compelling reason to trust the Bible is its radical transparency. If the Bible were simply a tool to promote a religious agenda, its authors would likely have omitted or softened details that reflect poorly on its central figures. Yet, time and again, the Bible presents its heroes not as flawless icons, but as deeply human.

Take King David, one of Israel's most celebrated leaders. He is portrayed not just as a warrior and poet, but as a man who committed adultery with Bathsheba and orchestrated the death of her husband, Uriah (2 Samuel 11). Such an unflattering account makes little sense if the goal were to promote an idealized version of Israel's leaders.

Similarly, the Bible records the repeated failures of the Israelites—including their worship of the golden calf at the very foot of Mount Sinai (Exodus 32). These stories, which could have been conveniently edited out or reframed, instead reveal a commitment to truth over image.

The New Testament continues this pattern of transparency. The Gospels openly portray the disciples—Jesus' closest followers—in moments of doubt, fear, and failure. Peter is shown denying Jesus three times (Luke 22:54–62), and Thomas famously refused to believe in the resurrection until he saw physical proof (John 20:24–29). These unflattering accounts would be unlikely inclusions if the writers were simply trying to promote a flawless image of their movement's founders.

This kind of unflinching honesty strengthens the Bible's credibility—especially in a world where historical revision and bias are common. It suggests that the biblical authors were more committed to preserving what actually happened than to crafting an idealized narrative. In fact, the Bible's willingness to expose weakness, failure, and even rebellion gives it a kind of moral and historical authenticity that few ancient texts can match.

## A Call to Respond

After weighing the evidence—historical, archaeological, manuscript, and prophetic—even some skeptics come to a surprising conclusion: the Bible may not only be historically credible, but spiritually true as well.

Its remarkable preservation, fulfilled prophecies, and transformative power point to more than just a human book. They suggest something deeper—something divine.

Of course, doubt and questioning are natural. But the Bible doesn't shy away from scrutiny; it has withstood centuries of it. The evidence doesn't demand blind faith—it invites honest exploration.

If the Bible truly is what it claims to be—God's word to humanity—then its message deserves more than intellectual curiosity. It calls for thoughtful reflection and, ultimately, a response. Not just a response of the mind, but of the heart.

The Bible doesn't merely ask for agreement—*it calls for transformation.*

Now that we've established a foundation for trusting the Bible, we can turn to one of its most debated narratives—a story many dismiss as myth, yet one supported by both Scripture and science. In the next chapter, we'll explore ***The Case for a Global Flood*** and see how evidence from around the world continues to affirm this ancient account.

# The Case for a Global Flood

*They say it rained—nothing vast—*

*Just soaked a ridge, then quickly passed.*

*But fossils found on every peak,*

*Reveal a flood—a massive sweep.*

*So grab your snorkel and take a ride—*

*This Flood was global, far and wide.*

## A Flood Anchored in Evidence

The idea of a worldwide flood is not just a matter of faith—it is supported by a growing body of geological, archaeological, literary, and biblical evidence.[1] Even long-time skeptics have begun reconsidering their assumptions in light of recent findings.

## Flood Legends Worldwide

One compelling line of evidence is the remarkable number of flood traditions preserved across cultures. From Asia to the Americas, the Middle East to the Pacific Islands, civilizations with no known contact recorded stories that follow strikingly similar patterns.

Geologists including **Dr. Andrew Snelling** have documented **thousands** of such legends in roughly **seventy** languages.[2] Many contain details closely paralleling the biblical account—not vague tales of seasonal flooding, but memories of a catastrophe far larger than any local disaster.

**Dr. John Morris** analyzed more than **two hundred** flood accounts and found striking consistency:

- 88% describe a favored family.
- 66% include a divine warning.
- 66% cite moral corruption or divine judgment.
- 95% specify that the catastrophe was a flood.
- 95% describe it as global in scale.
- 70% mention survival by a boat.
- 67% include animals being saved.
- 57% place the landing on a mountain.[3]

Across continents and centuries, the same narrative emerges: a family warned of divine judgment, a worldwide flood sent to confront human wickedness, the construction of a vessel, preservation of life, and a final landing on high ground. These similarities are too specific and widespread to dismiss as coincidence.

## Addressing Skeptical Objections

Doubters sometimes argue that these legends arose independently as metaphors or that they all stemmed from a local disaster. Yet this explanation raises its own problems. If all the stories were precisely the same, critics would claim collusion. Because they are not identical—but impressively similar—they dismiss them as coincidence.

A more reasonable conclusion is this: humanity preserved a shared memory of a single global catastrophe, adapted through different languages and cultures. The worldwide footprint of these legends points to a historical event, remembered in diverse forms yet unmistakably connected.

These widespread memories also appear in some of the world's oldest written texts, strengthening the case that they reflect a shared historical event rather than isolated cultural invention.

## Echoes in Ancient Literature

One of the most well-known examples is the Epic of Gilgamesh, which recounts a righteous man who survives a devastating flood by building a great vessel, which lands on Mount Nisir in modern-day Iraq, near Mount Ararat. Though details vary, the core story mirrors Genesis.[4]

## What the Bible Really Says

Some claim that the biblical Flood was merely local. Yet the Hebrew word for "flood"—*mabbul*—appears only in reference to Noah's Flood, not ordinary, regional floods.

The New Testament reinforces this understanding: the Greek word *kataklusmos* used in Matthew 24:39 and 2 Peter 3:6, literally means "cataclysm" or "overwhelming flood"—words describing a universal disaster.[5]

## Biblical Testimony and Purpose

Jesus affirmed the Flood's historicity:

> *Just as it was in the days of Noah, so also will it be in the days of the Son of Man. People were eating, drinking, marrying and being given in marriage up to the day Noah entered the ark. Then the flood came and destroyed them all. — Luke 17:26-27*

Peter echoes this truth:

> *By these waters also the world of that time was deluged and destroyed. — 2 Peter 3:5-6*

Genesis explains that humanity had become thoroughly corrupt, yet God preserved Noah and his family as a righteous remnant (Genesis 6:5-8). The Flood was therefore a real, global event combining divine judgment with the preservation of life through Noah.

> *The Lord saw how great the wickedness of the human race had become on the earth, and that every inclination of the thoughts of the human heart was only*

*evil all the time... But Noah found favor in the eyes of the Lord. — Genesis 6:5-8*

God gave Noah detailed instructions for building the ark (Genesis 6:13–21). The Flood was purposeful—both an act of judgment and an act of preservation.

## How the Flood Happened: Mechanisms and Models

Genesis emphasizes a dual source of water:

*In the six hundredth year of Noah's life, on the seventeenth day of the second month—on that day all the springs of the great deep burst forth, and the floodgates of the heavens were opened... — Genesis 7:11-12*

The "springs of the great deep" point to massive underground water reservoirs that had been held in place by the Earth's crust until they were suddenly released with explosive force.

Some researchers have explored how such an event could occur naturally. **Dr. Walt Brown's Hydroplate Theory** offers a detailed explanation of how subterranean water pressure, crustal rupture, and rapid geological activity might have interacted during a global Flood.

While not universally accepted, the theory provides potential insights into puzzling features: extensive sedimentary layers, widespread fossil burial, and major landforms such as the Grand Canyon.

From a biblical perspective, these processes were purposeful. God guided the preservation of life within the Ark and orchestrated the conditions that reshaped the Earth, integrating physical upheaval with moral and redemptive intent.

## Foundations: Pre-Flood Earth

To understand the Flood's physical possibility, consider pre-Flood geography. Dr. Brown proposes that roughly half of today's ocean water was trapped in vast chambers about ten miles beneath the Earth's surface. Granite caps held this water under immense pressure, forming a colossal underground hydraulic system ready to erupt.

Before the Flood, the continents formed a single landmass above what is now the Atlantic Ocean. Shallow seas and modest mountains dotted this land. This world remained relatively stable for centuries.

Then, the fountains of the great deep erupted, unleashing enormous geological forces that caused faulting, mountain folding, canyon carving, and other planetary-scale changes—laying the groundwork for the post-Flood Ice Age.

## Four Phases of the Hydroplate Theory

The sequence unfolds in four distinct phases:

1. Rupture Phase
2. Flood Phase
3. Continental Drift Phase
4. Recovery Phase

## Phase 1: The Rupture Phase—Catastrophic Beginnings

This was a moment of unprecedented upheaval—truly a point where *"all hell broke loose."* The Mid-Atlantic Ridge, visible today as a 10,000-mile underwater mountain chain from the Arctic to Antarctica, marks the scar of that rupture.

Over centuries, tidal forces heated and pressurized massive subterranean reservoirs. Eventually, the crust could no longer contain the buildup. It split open with unstoppable force, and **supercritical water**—a fluid in which the boundary between liquid and gas vanishes—erupted from the Earth at supersonic speed.

Some of this water shot skyward, mixing with a mist-filled pre-Flood atmosphere to produce torrential rain. Other portions froze into hail, snow and muddy ice, burying plants and animals. This sudden entombment may help explain phenomena such as frozen woolly mammoths with undigested vegetation still in their stomachs. It also contributed to the massive ice buildup that later set the stage for a post-Flood Ice Age.

According to this model, some debris may have escaped Earth's gravity entirely, potentially forming comets, asteroids, and meteoroids.

## Phase 2: The Flood Phase

The rupture unleashed torrents that reshaped the Earth's surface. Towering cliffs collapsed, and huge debris flows swept across the continents, burying life beneath layers of sediment at incredible speed. Fossils and strata formed rapidly—far faster than conventional models suggest.

Floodwaters stripped forests, leveled terrain, and buried organic matter under immense heat and pressure. This process formed coal and oil in a fraction of the time typically proposed. Vast, flat sedimentary layers puzzled geologists for decades, but the Hydroplate Theory offers a catastrophic explanation: these layers formed rapidly through global water movement.

## Phase 3: The Continental Drift Phase

Following the Flood Phase, the Earth's ruptured crust continued to separate, and the redistribution of surface weight caused the underlying crust to rebound upward. This led to the rapid rise of the Mid-Atlantic Ridge and the Atlantic Ocean floor by nearly ten miles, while the Pacific Ocean basin deepened and expanded.

The massive hydroplates then slid downhill, accelerating continental movement. These colossal shifts didn't just reshape the ocean floors—they lifted mountains, buckled continents, and rewrote the planet's topography in a matter of months or years.

As the plates collided, the crust crumpled and folded, forming mountain ranges such as the Himalayas and Rockies, carving deep ocean trenches, and giving rise to the fiery chain of volcanoes known today as the "Ring of Fire."

## Explaining Folded Mountain Ranges

Many granite mountain ranges appear bent and folded, defying slow-and-steady explanations.

The Hydroplate Theory proposes that this occurred while the rock was still malleable during rapid crustal movement—later cooling into the hard formations we see today.

## Phase 4: The Recovery Phase

After continental movement slowed, the crust stabilized. Lava continued to erupt along mid-ocean ridges, filling new basins and sealing fractures. As floodwaters receded, they pooled along continental shelves, creating the older shorelines and shaping modern coastlines.

Residual geothermal heat, combined with volcanic activity and atmospheric ash, intensified a brief but impactful Ice Age. Glaciers expanded, carving valleys, river systems, and other landforms. As the ice melted, rising oceans submerged low-lying regions and land bridges that once connected continents.

After Noah and the animals disembarked from the Ark on the mountains of Ararat, life began spreading outward across the continents. Temporary coastal plains, valleys, and exposed land corridors allowed migration to new regions. Birds, flying insects, and drifting seeds dispersed life even farther. As sea levels rose, these pathways disappeared, isolating populations and shaping global biodiversity.

In this way, life repopulated the Earth from a central refuge—fully consistent with both Scripture and the Hydroplate model's depiction of rapid, planet-wide geological transformation.

## Geological Clues for the Hydroplate Theory

If a global flood truly occurred, the Earth itself should bear its scars. According to this model, it does. From vast canyons and rapidly deposited sediment layers to frozen mammoths and upright trees fossilized through multiple strata, the planet shows evidence of catastrophe rather than slow, gradual geologic processes.

Here are three compelling lines of geological evidence that support the theory: the Grand Canyon, the eruption of Mount St. Helens, and the mystery of polystrate fossils.

## The Grand Canyon: Testimony of Catastrophe

The Grand Canyon stretches over 277 miles, plunges over a mile deep, and spans up to 18 miles across—a dramatic scar in the Earth's crust.

Yet the Colorado River, winding at its base, seems far too small to have carved such a massive feature. This raises an obvious question: *how did such a gigantic canyon form?*

Mainstream geologists say it took millions of years of slow erosion, but the Hydroplate Theory offers a radically different explanation. Rapidly retreating floodwaters carved the canyon after a global catastrophe—a massive, high-energy event, not the slow work of a trickling stream.[6]

## Sediment Layers that Defy Time

Before the canyon was carved, floodwaters deposited massive layers of mud, rock and organic material in rapid succession. The layers are remarkably flat and uniform, with virtually no evidence of erosion between them. This strongly indicates rapid deposition, not the slow, gradual processes often assumed by conventional geology.

This demonstrates that these strata were not exposed to long periods of weathering or biological disturbance, but were instead deposited in swift, continuous succession by a singular, cataclysmic event.[7]

As waters surged, they stripped vegetation and buried plants and animals under thick sediments. Many organisms were rapidly fossilized, providing a strong indication of sudden, high-energy deposition. Could these really be anything other than the *fingerprints of catastrophe?*[8]

## The Carving of the Canyon

Once the floodwaters began to recede, they did so with tremendous force. According to the Hydroplate Theory, the receding waters cut deep, fast-moving channels through the soft sediment layers.

The result? A massive canyon carved not over eons, but possibly within days or weeks. The Grand Canyon's smooth rock walls and sharp layer boundaries are exactly what you'd expect from a high-energy, short-duration event—not slow, gradual erosion.[9]

## The Great Unconformity: A Global Boundary Etched in Stone

The Great Unconformity is a dramatic boundary where old, tilted Precambrian rocks abruptly meet flat, younger sedimentary layers. It is as if two vastly different chapters of Earth's history were suddenly glued together, leaving out everything in between.[10]

Creation scientists propose this unconformity formed during a cataclysmic compression event triggered by the global flood. Subterranean waters bursting forth caused the crust to buckle and fracture. Older rock layers were tilted and broken, then vast sheets of water-saturated sediment slid across, smoothing and burying the rubble beneath fresh layers.

This global boundary, visible on multiple continents, fits a catastrophic flood model far better than a slow-and-steady geological timeline.

Imagine a force capable of bulldozing mountain ranges, tilting those massive rock layers steeply—as seen in the *Great Unconformity*—and burying the rubble beneath a fresh layer of mud and gravel. That's the scale and force of what we're talking about.

The Great Unconformity is more than a curious geological oddity. It is a global fingerprint of destruction and rapid change—a silent witness to a world once overwhelmed by water.

## The Isis Temple Anomaly: A Rock that Shouldn't Exist

Deep within the Grand Canyon lies a puzzling mystery known as the *Isis Temple Anomaly*—a hard, solid rock embedded within soft, uncemented sediment.

This shouldn't be possible if those surrounding sediments took millions of years to harden, as conventional geology claims.

If the layers around the rock were slowly compressed and cemented over eons, the rock would be locked into hardened stone.

**But it isn't.**

Instead, this anomaly sits like a boulder dropped into wet cement that never set.

Its very presence tells a different story—a rapid burial in soft, water-laid sediments that never had time to lithify fully before being uncovered again.[11]

This anomaly further supports catastrophic flood geology, highlighting sudden deposition rather than slow, uniform processes. Readers are encouraged to search for photos of the *Isis Temple Anomaly* online to fully appreciate this striking evidence.

In short, the *Isis Temple Anomaly* isn't just an oddity—it whispers of sudden upheaval, not slow formation.

## A Native Voice of Witness

The Havasupai people have passed down stories describing a great flood that once covered the Earth, mirroring aspects of the biblical account. Ancient oral traditions like these provide culturally independent testimony supporting a catastrophic origin.[12]

## The Canyon That Testifies

From the massive, flat sediment layers to the sharp geological boundaries—and even the traditions of local peoples—the Grand Canyon testifies to a catastrophic event.

The Hydroplate Theory doesn't just align with Scripture; it provides a scientifically detailed, visually compelling framework for understanding Earth's past.

Rather than a slow trickle of erosion, the evidence points to a sudden, global upheaval—the receding waters of a worldwide flood.

For a visual demonstration, see the YouTube video "Hydroplate Theory: Origins of the Grand Canyon" by Bryan Nickel.[13]

## Mount St. Helens: A Modern-Day Catastrophe that Rewrote the Textbooks

The eruption of Mount St. Helens proved to be a real game changer, offering scientists a front-row seat to rapid geological processes once thought impossible.

## Geology in Real Time

When the volcano erupted in 1980, it did more than just blow its top—it shattered assumptions about how the Earth's surface changes over time. In just hours, this modern catastrophe delivered undeniable proof that massive geological transformations can occur almost overnight, not over eons as traditionally believed.[14]

## The "Little Grand Canyon": Carved in Days

Dr. Walt Brown points to Mount St. Helens as a real-world demonstration of catastrophic geology in action. In moments, valleys were gouged out, sedimentary layers were stacked like pancakes, and entire landscapes were reshaped. Among the most eye-opening formations was the creation of the "Little Grand Canyon of the Toutle River"—a canyon 1,000 feet wide and 140 feet deep, carved not in millennia, but in just days.[15]

*Let that sink in.*

## A Blow to Uniformitarianism

Had a geologist stumbled upon this canyon without knowing its origin, they might've confidently declared it the product of millions of years of erosion. Yet it formed almost instantly. This single event is a direct challenge to the bedrock of uniformitarianism.[16] What happened next would drive the point home with undeniable force.

## The Power of a Single Catastrophe

The eruption on May 18, 1980, unleashed the largest landslide in recorded history—racing down the mountain at over 150 miles per hour and traveling 14 miles, obliterating everything in its path. Forests were flattened—rivers rerouted—roads and bridges vanished.

The explosion launched more than a cubic mile of rock, ash, and debris into the sky.[17] When the ash finally settled, enough sediment had fallen to bury Washington, D.C. in 14 feet of debris.[18]

But the surprises didn't stop there. In 1982, just two years later, melting snow triggered another catastrophic mudflow—this one carving out an intricate canyon system through the newly laid soft sediments **in only nine hours.**

The result was a canyon so detailed and layered, it mimicked formations thought to take vast ages to develop.

This wasn't just theory—it was geological change unfolding before scientists' eyes. The strata left behind by Mount St. Helens mirror ancient rock formations worldwide—features that conventional geology claims took millions of years to form. This undeniable similarity shows that under catastrophic conditions, layering happens fast.

In mere moments, nature rewrote the geological rulebook. What scientists once assumed required eons was instead accomplished in the span of a single day—a humbling reminder of how swiftly the Earth's surface can be transformed.

## A Living Laboratory for Catastrophic Geology

Mount St. Helens has become a living laboratory for creationists and flood geologists. It shows that rapid deposition, deep carving, and large-scale reshaping aren't just possible— they're observable.

Mount St. Helens stands as a testament—a modern, event that strengthens the credibility of the global flood model. It demonstrates that catastrophic processes can produce large-scale features like those seen at the Grand Canyon.

What took place at Mount St. Helens rattled the foundations of old-earth assumptions. The "Little Grand Canyon" carved in hours stands as a monument to catastrophe—a modern echo of the kind of global upheaval described in Genesis.

And the evidence doesn't stop there. Another amazing clue *stands upright*—literally.

## Polystrate Fossils: Evidence for a Global Flood

These towering fossils stand as silent witnesses to a sudden, world-altering catastrophe.

## Trees That Shouldn't Be There

Imagine a tree standing tall—not in a forest, but fossilized upright, its trunk piercing through dozens of feet of rock layers. This isn't science fiction—it's a real geological puzzle that points unmistakably to sudden catastrophe.[19]

## A Global Pattern, Not a Local Fluke

These vertical fossils—known as polystrate fossils—are found all over the world. They consist of tree trunks or plants that cut through multiple layers of sedimentary rock, often standing upright, frozen in time. They aren't rare anomalies—they're widespread, recurring, and incompatible with the uniformitarian timeline.

According to the slow, gradualist model of geology, each of these layers would have taken thousands—or even millions—of years to form. If that were true, the trees would have long since rotted away. Instead, these fossils speak of rapid burial—exactly the kind that would occur during a massive, global flood.[20]

Take the famous site in Joggins, Nova Scotia, for example. There, upright fossilized trees pass directly through layers of coal, shale, and sandstone. Not only are the trees preserved, but in some cases, the roots are still intact. There's no sign of slow decay or erosion between layers—just sudden burial under wave after wave of sediment.

And it's not just Nova Scotia. Polystrate fossils have been discovered in Tennessee, Kentucky, Germany, Australia, and beyond—testifying to a worldwide event, not a regional anomaly.[21]

## Buried Fast, Not Over Millennia

What makes these fossils extraordinary is what they don't show—any sign of erosion, decay, or weathering. The sediment wasn't trickling down over millennia; it was dumped rapidly, preserving trees before they had a chance to rot, collapse or be eaten by scavengers.

The sheer number and global distribution of these fossils strengthen the case for sudden, widespread burial on a catastrophic scale.

In short, polystrate fossils defy the idea of slow geological change. They stand as vertical witnesses to a moment in Earth's history when catastrophe, not calm, was the rule.

These layered trees don't just pierce stone—they pierce the uniformitarian narrative, pointing unmistakably toward the kind of rapid, global upheaval described in Genesis.

## Other Evidence for a Global Flood:

## Fossilized Clams on Mountain Tops

What do clams, mountaintops, volcanoes, and misplaced fossils all have in common? They testify to a catastrophic global event—one that reshaped Earth's surface in weeks, not millennia.

Consider a surprising discovery: fossilized clams found high in the mountains—not just fossilized, but *clamped shut*. Normally, when a clam dies, its muscles relax and the shell opens. These clams, however, were buried alive, quickly and deeply.[22]

Only an event of unimaginable scale could trap marine life on mountaintops before decay or shell opening—exactly what a global flood could produce.

## Marine Life Far from the Ocean

These are not rare, isolated finds. Fossilized corals, seashells, and even marine reptiles have been discovered at high elevations across the globe—from the Himalayas to the Andes—far from today's oceans. Their presence indicates that vast regions were once submerged.

The scale, distribution, and preservation of these fossils cannot be explained by ancient inland seas or tectonic uplift over millions of years. Delicate organisms would have been crushed, scattered, or weathered beyond recognition. Instead, they are intact and widespread—a signature of calamity and sudden burial.

*Something fast. Something global. Something unmistakably flood-like.*

## Layers of Lava and Sediment: A Chaotic Tapestry

Throughout the rock record, layers of volcanic lava flows interwoven with water-deposited sediments, forming a geological tapestry. What could cause alternating layers of ash, lava, and mud in such a short vertical span?

During a global flood, volcanic eruptions would have burst forth violently as the Earth's crust fractured, while floodwaters surged and receded in massive waves.[23] Each surge brought sediment, and each eruption deposited ash and lava—stacking layers in a chaotic pattern only catastrophe could produce.

## A Radiocarbon Spike: The Global Die-Off

Laboratories worldwide have identified an abrupt and dramatic clustering of radiocarbon "death dates" centered around 5,000 years ago. Fossilized wood, bone, peat, and shells from across continents reveal the same pattern: a sudden, simultaneous collapse of life on a global scale.

This is not conjecture; it's quantifiable, repeatable evidence—a radiocarbon signature that defies the slow, uniform geological expectations. Instead of gradual extinction, the data indicate a single, cataclysmic event that nearly wiped out life instantaneously. The timing and scale of this radiocarbon spike align closely with the biblical account of a worldwide flood—a moment when the natural order was violently interrupted and the Earth itself was reset.[24]

## Fossils Out of Order: Disrupting the Geologic Column

Perhaps most unsettling for traditional geology is the presence of fossils "out of order" in the geologic column.[25] Marine fossils sometimes appear above terrestrial layers, along with other misplacements, undermining the tidy timelines proposed by uniformitarian models. These observations support the idea of turbulent, high-energy deposition, where sediment, organisms, and debris were violently mixed and rapidly buried.

Taken together, these lines of evidence point to a single explanation: a global upheaval that lifted seas onto mountains, buried life in moments, and reshaped the planet. This is not speculation—it is the story etched into the Earth's crust, demanding catastrophic interpretation.

## Additional Biblical Evidence for the Global Flood:

While historical, archaeological, and manuscript evidence help confirm the Bible's reliability, the most powerful case for a worldwide event comes from Scripture itself. The Flood isn't an isolated story tucked away in the pages of Genesis—it's woven into the theological fabric of the entire Bible, echoed in genealogies, covenants, teachings of Jesus, and the symbolism of salvation. Let's examine five key pillars of biblical support:

## 1. Genealogies and the Universality of the Flood

One of the clearest signs that the Flood was not a localized event comes from the genealogies in Genesis. The unbroken lineage from Adam to Noah in Genesis 5—and from Noah's sons to all post-flood nations in Genesis 10—underscores a single, continuous human family tree.

If the upheaval had only affected one region, we would expect to see genealogical or historical traces of unaffected human populations elsewhere. Instead, Scripture presents a stark reality: Noah's family was the only one spared, and from them, the entire earth was repopulated.

This narrative of a global restart doesn't suggest a regional disaster—it demands a worldwide event.

## 2. The Ark: Built for Survival on a Global Scale

The Ark's design reinforces the universality of the Flood. According to Genesis 6:14–16, the Ark measured approximately 450 feet long, 75 feet wide, and 45 feet high—dimensions fitting not for a river raft, but for a vessel capable of surviving a planet-wide calamity.

Noah didn't simply gather animals from a nearby valley; he brought representatives of every land-dwelling, air-breathing creature. Such extensive preparation only makes sense if the entire planet was at risk.

If the flood had been local, escape would have been as simple as moving to higher ground. Scripture makes it clear: *there was nowhere to run.*

## 3. Affirmation of the Flood's Global Scope

The universality of the Flood is woven throughout Scripture. Jesus compared coming judgment to the days of Noah (Luke 17:26-27), emphasizing the suddenness and scope of the event. Similarly, the apostles describe the Flood as a defining example of God's global judgment and preservation of the faithful. Rather than quoting verses already presented, this section focuses on the theological and symbolic significance of the Flood as a worldwide event.

## 4. Baptism: A Symbol of Judgment and Redemption

Peter draws a striking parallel between Noah's Flood and baptism: *"This water symbolizes baptism that now saves you also… through the resurrection of Jesus Christ."* (1 Peter 3:20-21)

Baptism, the outward sign of salvation, is directly linked to the waters of Noah's day. Just as the Flood washed away a sinful world, baptism represents both the cleansing of sin and the beginning of a new life.

This symbolism only holds its full weight if the judgment was universal. A local event could not serve as a parallel to God's global act of redemption through Christ. What occurred in Noah's time was more than judgment—it was a divine reset, just as baptism represents a spiritual rebirth.

## 5. God's Covenant with All Creation

After the waters receded, God made a covenant not just with Noah—but with all living creatures:

> *Never again will all life be destroyed by the waters of a flood…* —Genesis 9:11

To seal the promise, He gave the rainbow as a universal sign of this commitment.

If the flood had only impacted a small region, a global promise would seem unnecessary. Instead, the scope of God's covenant matches the scope of the event: worldwide destruction followed by worldwide assurance.

This promise reminds us that this incident was no ordinary disaster. It was a singular, world-altering act of judgment—and mercy.

## Conclusion: The Evidence Speaks

The case for a global flood is not built on myth or speculation—it rests on converging geological, archaeological, cultural, and biblical evidence. From fossilized marine life found on mountains to flood legends preserved worldwide, and from the precise dimensions of Noah's Ark to the dramatic features of the Grand Canyon, the evidence points to a sudden, planet-wide deluge.

Geological features that defy gradualism—like polystrate fossils, the Great Unconformity, and the rapid canyon formation observed at Mount St. Helens—challenge the assumption that Earth's features all formed slowly over eons. Instead, they tell the story of a catastrophic, water-driven transformation consistent with the biblical account.

The Bible reinforces this understanding—not as a regional flood, but as a world-resetting act of divine judgment. As discussed, Jesus and the apostles affirm the Flood as a historical, global event, and the genealogies, Ark instructions, and covenantal promises underscore its universality and purpose. Even the symbolism of baptism—death, cleansing, and rebirth—traces its origin to the waters of Noah's day.

Ironically, the secular scientific community readily accepts ancient oceans on Mars—a cold, arid planet with no liquid water today—while dismissing the possibility of a global deluge on Earth, a planet already covered in water. This double standard doesn't reflect a lack of evidence, but philosophical divide between naturalistic assumptions and a worldview open to divine intervention.

Ultimately, the Flood narrative is not just about fossils or canyons—it's about understanding Earth's past through the lens of judgment, mercy, and redemption. The evidence stands open, waiting for honest inquiry and reflection. Whether viewed through the Hydroplate Theory or another catastrophic model, the conclusion is clear: ***Earth is telling a story—and it aligns with the account in Genesis.***

# A Universe Too Perfect to Be an Accident

*They say the universe just got lucky—true.*

*Gravity's perfect, and the sky's still blue.*

*Was it chance that set the cosmos in sync,*

*Or is someone upstairs with a crafty wink?*

## Introduction: A Case for Intelligent Design

How did everything get so perfect?

From the force of gravity to the structure of atoms, the universe operates with astonishing precision. A tiny shift in any one of dozens of physical constants—and life as we know it would cease to exist.

Many scientists acknowledge that this level of meticulousness is too exact to dismiss. This idea is known as the ***Anthropic Principle***—the recognition that the universe's physical laws and constants are so precisely calibrated to support life that we can observe them only because they allow observers like us to exist.[1]

While some see mere coincidence, the evidence points far more convincingly to the deliberate work of an extremely sophisticated mind.

Consider Mount Rushmore. Would anyone seriously suggest that George Washington's face, along with those other presidents, randomly formed on that mountain from natural weathering? Instead, we immediately recognize it as the product of immense effort and an intelligent creator.

Yet, when faced with a universe—or a human being like George Washington himself—vastly more complex than a lifeless sculpture, we are often told it simply appeared by chance.

Here's the real question: Is that fine-tuning just a cosmic accident—or does it point to a brilliant intellect behind it all?

In this chapter, we'll explore the finely tuned universe, compare intelligent design with evolutionary theory, and examine what the evidence suggests about our origins.

## The Fine-Tuned Universe: The Fingerprints of God

From a young Earth, non-evolutionary perspective, the universe's fine-tuning isn't just a lucky arrangement of natural laws that happened to support life over billions of years. Instead, it points to a deliberate blueprint—an intelligently crafted cosmos designed from the beginning to support life.

Consider these remarkable examples of this extraordinary precision:

## 1. The Cosmological Constant:

The cosmological constant, a term in Albert Einstein's equations of general relativity, represents the energy density of empty space—what we call dark energy. It plays a critical role in how fast the universe expands.

If this constant were even slightly larger, the universe would have expanded too rapidly for galaxies, stars, and planets to form. If it were slightly smaller, gravity would have pulled everything back together too quickly. Either way, life as we know it could never exist.

This razor-thin margin points not to random chance—but to a purposeful, intelligent designer who fine-tuned the cosmos from the very start.[2]

## 2. The Strength of Gravity:

Gravity is the force that holds stars, planets, and galaxies together—and it had to be *just right*. From the initial creation event, gravity played a crucial role in balancing the outward expansion of the universe with the inward pull that draws matter together.

If gravity were even slightly weaker, stars would never form, and planets like Earth wouldn't exist. If it were slightly stronger, stars would burn through their fuel too quickly, dying out long before life could emerge.

This exact calibration enables the formation of long-lived, stable stars like our Sun—an essential prerequisite for life. From an intelligent design perspective, such fine-tuning suggests not a random accident, but an intentional setup—a universe configured from its inception to be life-sustaining.

## 3. The Electromagnetic Force:

The electromagnetic force governs the interactions between charged particles and is essential for the stability of atoms and molecules. It determines how tightly electrons bind to atomic nuclei and how atoms connect to form molecules—such as water, proteins, and DNA.

If this force were even slightly stronger or weaker, atoms wouldn't form stable structures. Without stable atoms, complex chemistry—and life as we know it—would be impossible.

That this force is so precisely calibrated to permit the existence of complex molecules points unmistakably to deliberate design. In a purely random existence, such precision would be extremely improbable. Yet, in a universe created on purpose, it reflects the deliberate hand of a Creator who set in motion the fundamental forces to support life from the outset.

## 4. The Ratio of Proton to Electron Mass:

The mass of a proton is approximately 1836 times that of an electron. This specific ratio is crucial for atomic stability—it allows electrons to orbit nuclei in a way that forms stable atoms and, ultimately, the chemical bonds that make up all matter.

If this ratio were even slightly different, atoms would become unstable, and essential molecules—like those in DNA and proteins—couldn't exist. Life would be chemically impossible.

Such a finely tuned relationship defies the odds of random chance.[3] In a universe without purpose, this precise balance is incredibly difficult to explain. Conversely, if the universe were deliberately designed to support life, the exactness of this ratio makes perfect sense, pointing to a Designer who structured the cosmos as a habitat for life from the very beginning.

## 5. The Strong Nuclear Force:

The strong nuclear force is what binds protons and neutrons together in the nuclei of atoms. Without it, atomic nuclei would fly apart, and the essential elements for life—like carbon, oxygen, and nitrogen—couldn't exist.

This force must be finely tuned. If it were just slightly weaker, atoms wouldn't hold together. If it were slightly stronger, the fusion processes inside stars couldn't occur properly, preventing the formation of the heavier elements needed for life. Either way, a life-supporting universe would be impossible.

This exact balance isn't easy to explain by random processes. It suggests that the strong nuclear force was deliberately calibrated to make life possible—a signature of intelligent design embedded in the very laws of physics.

## 6. The Universe's Expansion Rate:

In the moments immediately following the creation event, the rate at which the universe expanded was critical. Observations of redshift in light from distant galaxies confirm this expansion, showing that the universe is continually stretching. If the expansion had been even slightly faster, matter wouldn't have clumped together to form stars and galaxies. If it had been slower, the universe would have collapsed in on itself before stars or planets could ever form.

Instead, the universe expanded at precisely the right rate—allowing it to cool at just the right pace, galaxies to take shape, and stars to ignite. This razor-thin margin for error and exact calibration doesn't look like luck; it strongly points toward an intelligently designed cosmos, carefully crafted from the foundational moments for life to thrive.

## The Human Mind: A Masterpiece of Design

Beyond the cosmic ballet of stars and the precise constants governing all existence, we find equally astonishing evidence of design in the biological realm, most notably within the human mind. Its capacity to process vast amounts of real-time data, make complex decisions, and adapt on the fly—as exemplified by something as common as driving a car—demonstrates a level of computational sophistication that rivals, or even surpasses, the most advanced systems, including theoretical quantum computers.

This isn't merely a collection of neurons; it is an intricate, adaptable, and self-aware system that points powerfully to an intelligent origin. The very existence of such a mind—capable not only of survival but of comprehending the structure of the universe—is itself a profound testament to a Creator.

Just as the vastness of the cosmos reveals precision on a grand scale, the natural world around us offers its own intricate examples of brilliant craftsmanship—fashioned with the same intentionality and care.

## Nature's Ingenious Designs: The Folding Patterns of Leaves

The incredible, pre-budding folding patterns of leaves are a stunning example of advanced engineering. These intricate folds, reminiscent of the art of origami, allow leaves to pack themselves efficiently into tight spaces. In fact, this pattern is among the most efficient packing arrangements known on Earth. NASA even adopted these leaf-folding patterns for the parachutes used in space capsule returns, demonstrating how nature's ingenuity inspires human innovation.

Such precise and purposeful design reflects the wisdom and foresight of a Creator who engineered nature's systems to function with elegance and efficiency. The intelligent planning evident in something as simple as a leaf mirrors the fine-tuned meticulousness of creation itself—a unified blueprint that spans from the microscopic to the galactic.[4]

Indeed, almost everything in nature looks miraculous—a testament to brilliant craftsmanship, evident in every delicate detail and grand cosmic structure. This profound level of intentional arrangement, from the smallest cell to the largest galaxy, speaks to a guiding intelligence rather than mere chance.

## The Structure of Spider Silk: A Material Beyond Compare

Spider silk is another astonishing marvel of natural engineering that continues to elude human replication, showcasing the brilliance of an intelligent Creator's handiwork. By weight, spider silk is up to five times stronger than steel—an extraordinary fact that underscores its unmatched design. Yet, it remains incredibly elastic, enabling spiders to build webs of extraordinary precision—perfectly crafted for capturing prey. The fibers are lightweight, stronger than most synthetic materials, and elastic enough to stretch and absorb the energy of incoming prey without breaking. The delicate balance of strength and flexibility is a testament to the wisdom behind its creation.

Scientists have long attempted to replicate spider silk synthetically, but no man-made material matches its unique combination of durability, lightness, and flexibility. This exceptional design found in nature highlights not only the genius of biological construction but also points to a fine-tuned system in which every aspect of an organism—from its physical body to its most intricate structures—serves a specific and purposeful function. The remarkable precision and engineering of spider silk reflect the intentionality and forethought of a Creator who meticulously fashioned the natural world to operate in harmony and perfection.

## The Honeycomb Structure: A Marvel of Efficiency

The honeycomb, crafted by the humble honeybee, is an incredible example of natural ingenuity. The hexagonal shape of each cell allows for the most efficient use of space—minimizing the amount of wax the bees must produce while maximizing storage capacity for honey and pollen. This structure is widely regarded as the world's most efficient space-filling arrangement possible in two-dimensional space, requiring the least amount of material for the greatest strength and volume. The perfection of the honeycomb—its symmetry, precision, and extraordinary efficiency—demonstrates an inherent understanding of geometry that far surpasses what even human engineers initially conceived.

Just as the honeycomb maximizes the utility of its resources, the entire ecosystem reflects a carefully orchestrated balance of interdependent systems designed to maximize life and purpose. It is a profound reflection of intelligent crafting, showcasing an efficiency that scientists and architects alike continue to admire.

## The Great Riddle: Chicken or Egg?

Here's a question that has puzzled minds for centuries: *What came first, the chicken or the egg?* If the egg came first, how could it exist without a chicken to lay it? But if the chicken came first, where did it come from if not an egg? Evolutionary theory struggles to provide a clear, satisfying answer to this age-old puzzle.

From a biblical perspective, however, the answer is straightforward: God created the first two chickens—one rooster and one hen—and from there, the hen laid many eggs. **Problem solved!**

While the chicken-and-egg dilemma highlights the circular problems in biological evolution, the same kind of paradox appears when we examine the universe itself.

## The Fine-Tuned Universe: An Evolutionary Conundrum?

Evolutionary theory proposes that life on Earth emerged over billions of years through natural selection acting on random mutations. However, if the cosmos is so precisely fine-tuned to support life, a pressing question arises: *Could evolution alone account for life's origins?*

1. *The "Lucky Accident" Fallacy:*

Evolutionary theory depends heavily on the idea that life emerged by chance. Yet the exact and highly specific conditions required for living organisms to exist render the likelihood of such an event not just incredibly low, but virtually impossible. The exact calibration of the universe points to a much more intentional process, suggesting the involvement of an intelligent designer rather than random chance.

2. *The Origin of Life:*

Evolution explains how life changes over time, but it does not adequately address how life began in the first place. The universe's precise fine-tuning strongly suggests that life could not have arisen at all without these perfect, pre-established conditions, once again pointing to an intelligent Creator rather than a random, gradual process emerging from an environment ill-suited for life.[5]

### 3. *The Complexity of Life:*

Even within a finely tuned creation, evolutionary theory struggles to explain how complex biological systems, such as cellular machinery, could emerge solely through random mutations. The meticulousness required to produce these systems suggests that evolution alone does not adequately account for life's complexity without purposeful input. This brings us back to the concept of intelligent design, where complexity is deliberately introduced, not accidental.

## Bold Confessions: When Science Confronts Faith

This paraphrased summary from Nobel Prize-winning biologist George Wald is worth noting:

> *When it comes to the origin of life, there are only two possibilities: creation or spontaneous generation. There is no third way. Spontaneous generation was disproved one hundred years ago, but that leads us to only one other conclusion: supernatural creation.*[6]

This statement highlights the intellectual tension many evolutionary scientists face. And Wald is not alone.

Professor **Louis Bounoure**, Director of Research at France's National Center of Scientific Research, famously stated:

*Evolution is a fairy tale for grown-ups. This theory has helped nothing in the progress of science. It is useless.*[7]

That's not the voice of a fringe religious apologist—that's the opinion of a leading figure in European science.

Similarly, **Dr. T. N. Tahmisian** of the U.S. Atomic Energy Commission issued a scathing indictment of evolutionary theory:

*Scientists who go about teaching that evolution is a fact of life are great con-men, and the story they are telling may be the greatest hoax ever.*[8]

Statements like these demonstrate that deep skepticism about evolution doesn't only come from religious circles. In fact, some of the harshest criticisms come from within the scientific community itself. **Dr. Robert Jastrow**, astronomer, physicist, and founder of NASA's Institute for Space Studies, summed it up this way:

*For the scientist who has lived by his faith in the power of reason, the story ends like a bad dream. He has scaled the mountains of ignorance; he is about to conquer the highest peak; as he pulls himself over the final rock, he is greeted by a band of theologians who have been sitting there for centuries.*[9]

# The Emperor's New Theory

# Why Evolution Still Reigns

*Some cling to a tale that's fraying and thin,*

*Not for the facts, but the comfort within.*

*For truth has a cost that few want to pay—*

*So they mask it in theories and turn truth away.*

## The Librarian and the Locked Door

There once was a brilliant librarian who spent her life in a vast, ancient library. This library held every book ever written about science, philosophy, and the natural world. She prided herself on knowing them all.

At the end of a quiet hallway stood a sturdy wooden door with no label—just a small plaque that read, *"Private: Do Not Enter."* She passed it every day but never gave it much thought.

One evening, while studying a rare volume on the origins of life, she noticed a footnote that read, *"For a fuller understanding, see the volumes behind the locked door."*

Curious, she found the key buried in the archives. It was old and rusted, but it fit. As she opened the door, she found a hidden room—filled with books she had never seen. Books that answered questions she didn't even know she had. They were authored not by human will, but divinely inspired by a higher intelligence. The ideas were unsettling, beautiful, and full of implications she wasn't ready for.

She closed the door. Locked it again. Hid the key.

When asked later why she didn't read the books, she replied, "Because if they're true... I'll have to change everything."

So she returned to her familiar shelves and continued teaching others the same theories she had always taught—because it was safer, simpler, and demanded nothing of her.

## Unmasking the Factors That Drive Belief in Evolution

Like the librarian in the story, many today often embrace Darwinian evolution—not because of overwhelming evidence, but because it's the familiar narrative—the one that demands the least personal cost.

Though evolution is widely presented as settled science, much of the commonly cited support lacks the empirical rigor needed to substantiate such a sweeping claim. An objective thinker—armed with comprehensive, unfiltered information—may find that the core assumptions of Darwinian evolution deserve serious scrutiny.

Yet our culture has gone to great lengths to sidestep this uncomfortable reality.

So the question remains: **why?**

## Perhaps the Reason Lies Deeper than Data

For many, belief in macroevolution isn't always grounded in hard evidence, but in a complex web of cultural conformity, incomplete education, personal comfort, professional pressure, pride, or even the desire for moral independence.

Sometimes, the motivation is as simple as not wanting to be wrong, or not wanting to lose the life one has built.

## When the Mind Shifts but the Heart Holds Back

I'm reminded of an old friend who once openly professed his disbelief in God. Though our views differed, I respected his right to follow his own convictions.

We eventually lost touch, but twenty years later, we reconnected around a bonfire. As we caught up, the topic of evolution came up—something we had never seriously discussed before.

To my surprise, he expressed genuine skepticism about the theory. His doubts, he explained, were rooted in scientific concerns. He admitted he found little compelling reason to fully embrace evolution as it's commonly presented.

I asked him, "If evolution isn't true, how can you still be an atheist?"

Without hesitation, he acknowledged that a higher power must exist—because without one, there's no real explanation for the existence of the universe.

While I was encouraged by this shift in his thinking, he quickly clarified that he still rejected the God of the Bible.

He had moved from atheism to deism—a belief in an impersonal creator who designed the universe but remains distant, unknowable, and uninvolved.

His story is a reminder that belief isn't always shaped by data—but by what we're willing to do with the implications of that data.

Even when the mind begins to change, the heart may resist. Accepting the possibility of a Creator isn't just an intellectual shift—it's a personal and moral one. And for many, that's the harder step to take.

Even so, I found it stunning how quickly he dismissed the biblical God—especially given the depth of evidence supporting the biblical worldview. His journey reminded me of something I've seen many times before:

*The barrier to belief is not always intellectual; it can be deeply personal.*

## The God We Want vs. the God Who Is

It's fascinating that someone like my friend would choose to believe in a god who is unknowable, detached, and unsupported by evidence—while simultaneously rejecting the biblical God, who is clearly defined, self-revealed, personally involved, and backed by compelling evidence.

Deism offers a convenient middle ground: a way to acknowledge a creator without submitting to the personal involvement or moral accountability that comes with the God of Scripture.

Unlike my forthright friend, many atheists and secular humanists continue to embrace evolution—even when the evidence is far from conclusive.

But this raises a deeper question:

*Could it be that the real barrier isn't scientific at all—but personal, moral, or philosophical?*

## When Science Becomes a System of Self-Preservation

Some individuals simply haven't been exposed to perspectives outside the dominant evolutionary narrative.

For them, belief in evolution isn't the result of careful investigation, but simply the story they were taught. These individuals deserve understanding rather than criticism.

Some people, however, may cling to evolution for reasons that run deeper. Some prefer a worldview that frees them from moral accountability.

Still others conform to the consensus because it's easier—and safer—than challenging the status quo.

Within the scientific community, belief in evolution can also be driven by professional incentives. Research funding, academic promotion, and peer recognition often favor those who support the prevailing theory.

To dissent is to risk ridicule, marginalization, or the dreaded label of "intellectually unsophisticated." In such a climate, even questionable ideas stick around—not because they are true, but because challenging them is too costly.

## Freedom from Meaning: What Evolution Really Offers

Beyond peer pressure or professional incentives, there are deeper reasons why some cling to evolution. For many, the appeal isn't scientific—it's profoundly personal.

Aldous Huxley, the influential writer and philosopher, admitted that his rejection of a meaningful universe was not driven by evidence, but by a desire for freedom from moral restraint:

> *I had motives for not wanting the world to have meaning... For myself, and most of my contemporaries,*

*the philosophy of meaninglessness was essentially an instrument of liberation. Liberation from a certain system of morality. We objected to the morality because it interfered with our sexual freedoms.*[1]

Huxley's honesty reveals a reality many prefer to keep hidden: personal desires often shape philosophical commitments.

For some, evolution offers a convenient escape from the implications of divine authority—a way to live without guilt or moral constraints. It's like trying to ignore an elephant in the room—undeniable, obvious, and impossible to miss, unless you're determined not to see it.

**THE EVOLUTIONIST'S DILEMMA**

The influence of worldview over science is echoed by Harvard geneticist Dr. Richard Lewontin. He acknowledged that many in the scientific community hold to materialism not because the data demands it, but because their philosophy does:

> *We have a prior commitment, a commitment to materialism… we cannot allow a Divine Foot in the door.*[2]

In other words, when belief begins with a commitment to exclude God, all evidence is filtered to fit that framework—regardless of where it naturally leads.

The challenge, then, isn't just to correct flawed science—*it's to reclaim science from the grip of secularism.*

Until that happens, truth will continue to bow to ideology rather than confronting it.

That's why the debate over evolution isn't just scientific. It's personal. It's philosophical.

It's about **autonomy.**

It's about **authority**.

And it's about what we're willing—or unwilling—to surrender.

Dr. John MacArthur puts it bluntly in *The Battle for the Beginning*:

> *It boils down to the sheer love of sin. People want to feel comfortable in their sin, and there is no way to do that without eliminating God. Get rid of God and you*

*erase all fear of the consequences of sin. So even though sheer irrationality is ultimately the only viable alternative to the God of Scripture, multitudes have opted for irrationality just so they could live guilt-free and shamelessly with their own sin. It is that simple.[3]*

In the end, evolution's staying power isn't rooted in textbooks or test tubes.

It's grounded in hearts that would rather choose illusion than face accountability.

As Romans 1:20 reminds us:

*For since the creation of the world God's invisible qualities—his eternal power and divine nature—have been clearly seen, being understood from what has been made, so that people are without excuse.*

The persistence of evolution in our culture is not just a scientific issue.

*It's a social, psychological, and spiritual phenomenon, best captured by a timeless tale.*

## The Emperor's New Clothes: Why Evolution Persists Without Evidence

The situation reminds me of the famous tale of *The Emperor's New Clothes*. In the story, the emperor, driven by vanity and desire to appear wise, is deceived into believing he is wearing magnificent clothes made of a magical fabric that is invisible to anyone "unfit for their position" or "too stupid to understand." His advisers, eager to please, go along with the lie, praising the emperor's "new clothes." The emperor, not wanting to seem foolish, also pretends to admire his outfit—even though he's walking around naked. The crowd, fearing ridicule for speaking the truth, plays along with the delusion—until a child innocently shouts, "But he isn't wearing anything at all!"

In much the same way, evolution has become the Emperor's "new clothes" of our time—not accepted for its solid scientific merit, but because many fear appearing ignorant or being ostracized.

Just as the emperor's subjects were afraid to admit the obvious, many today feel pressure to affirm evolution, even when the evidence seems weak or inconsistent.

But the truth is there—*clear to anyone with the courage to look.*

## Conclusion: The Battle Between Worldviews

Much like the emperor's deception, many cling to the belief in evolution—not because it's firmly grounded in irrefutable science, but because it allows them to avoid confronting deeper, often uncomfortable truths—especially the possibility of a Creator with moral authority over our lives.

To challenge this prevailing narrative can feel like questioning the emperor's wardrobe: socially risky and intellectually costly.

For some, evolution serves less as a scientific conclusion and more as a philosophical covering—an intellectual garment woven from speculation, consensus, and cultural momentum. But like the emperor's robes, this garment may look convincing until it's held up to scrutiny. And when it is, the fabric begins to unravel.

This is not just a scientific dispute. It's a clash of worldviews.

The real question isn't whether evolution is true—it's why it persists with such cultural force. Is it because the facts demand it, or because the alternative threatens our autonomy?

Before you move on, take a moment to reflect:

This chapter isn't just about evolution—it's about the layers beneath belief:

The stories we accept.

The truths we avoid.

The price we're willing to pay for independence.

If you've begun to sense that the issue runs deeper than data—that it touches on purpose, worldview, and moral accountability—then don't stop here.

We invite you to turn to Stumbling Block #16: *The Bible vs. Evolution: The Ultimate Collision*, where you'll discover a truth that offers not only answers, but hope.

There, you'll meet the Creator many have tried to forget—

*but who has never forgotten you.*

Truth doesn't vanish just because it's inconvenient—it waits to be rediscovered.

Don't settle for illusion.

Step beyond the veil—**into the light of truth, and the arms of the One who made you.**

# The Cultural Fallout of Evolution

*When truth is discarded and falsehoods accepted,*
*We reap the consequences just as expected.*

When societies exchange divine wisdom for human speculation, the effects ripple through every part of life—from our view of human dignity to the laws that shape our culture. The impact of evolution goes far beyond biology; it distorts how we see ourselves, our purpose, and the world around us.

This fictional allegory shows how replacing higher purpose with random chance can erode the very foundations of society.

## The Garden of Mirrors: An Allegory on Darwinian Thinking

There was once a beautiful kingdom called Arcadia, perched on the edge of a great cliff. Its people lived by the Book of the Builder, a sacred text passed down through generations. It taught that the land, the people, and the sky above had all been crafted by a wise and loving Architect. Because of this belief, the citizens honored one another, protected the weak, and lived with direction and hope.

At the center of the city stood a gleaming tower called the Hall of Origins. Inside was a great mirror that reflected not only the people's faces—but their identity. When they looked into it, they didn't just see skin and bone; they saw purpose, design, and divine worth. Children were taught from a young age: **"You were made. You matter. You're more than dust."**

One day, a traveler arrived from a distant land. He brought with him a new mirror—sleek, modern, and different. He called it the Mirror of Chance and claimed it told a deeper truth. "Look closely," he urged. "There is no Builder. No Architect. Only random forces, time, and survival."

Many scoffed. But curiosity crept in.

First the scholars entered the Hall and replaced the old mirror. Then came the teachers, the students, the judges, and eventually the leaders. When people looked into the Mirror of Chance, their reflection changed. They no longer saw a masterpiece—they saw an accident. A cosmic mistake. A collision of molecules.

Some were disturbed. Others felt liberated. Without a Creator, there were no rules, no judgment, and no higher calling. "We are free," they cried. "Free to make our own meaning!"

And so, they did.

Laws changed. The weak lost protection. The unborn, the elderly, the inconvenient—deemed expendable. Families fractured.

Trust eroded. Violence grew. The city that once stood tall began to crumble.

When the Architect sent messengers to warn them, they were mocked and silenced. "We have evolved beyond such tales," they declared.

Eventually, the cliff beneath the kingdom began to crack. The foundation, once anchored in design, eroded. A tremor shook the land. The tower fell. And the people—confused and broken—were left staring into a mirror that no longer reflected hope, only a shattered image, irrevocably lost.

Though fictional, Arcadia reflects a sobering reality we can't ignore. And yet—even in ruin—the possibility of redemption remains.

## The Fallout of Darwinism: When Speculation Masquerades as Fact

Darwin's theory of evolution—though widely accepted as scientific orthodoxy—rests on assumptions and interpretive leaps, not irrefutable evidence. Despite its confident claims, God's truth remains a *persistent thorn in evolution's shoe.*

The implications extend into every corner of culture. Darwinism functions as a belief system—one that strips God from the equation and replaces order with chaos. It reduces people to random byproducts of unguided processes, undermining the sacredness of life and paving the way for a culture marked by confusion, dehumanization, and spiritual drift.

So, how does this deception speak to our culture today?

Let's imagine it not just as a theory, but as a voice—scientific in tone, seductive in message.

I call her… the **Siren of Self**.

## The Siren of Self: A Personification of Evolutionary Deceit

*She stands at the crossroads, cloaked in reason but dripping with pride.*

*They call her Science, but her lips whisper deceit.*

*Her charm is not truth—it's imitation, well-dressed and well-spoken.*

*She scoffs at the old ways. Rolls her eyes at design.*

*And laughs at the thought of a Creator with a plan.*

*"Come," she says, "you are dust and nothing more.*

*No Judgement waits. No meaning binds. No soul endures."*

*Her sleek mirror flatters the ego but distorts the face.*

*Fools flock to her—not for truth, but for the freedom of lies.*

*She is not wisdom. She is rebellion, polished and praised.*

*She sits beside Hunam Autonomy and the throne of self-worship,*

*Her twin daughters—lovers of autonomy and enemies of light.*

*Together they build towers on sand,*

*Declare man god, and mock the heavens.*

*Yet the wind still comes. The foundation still breaks.*

*And when the dust settles, her followers stand empty—*

*Enlightened, perhaps, but lost.*

*They followed a voice… but it was not the Shepherd's.*

*They embraced a queen… but she wore no crown.*

*She was not truth. She was the echo of Eden's lie.*

*And her voice still whispers through our culture today.*

This poetic metaphor captures the seductive pull of evolutionary thinking—appealing, progressive, and self-flattering, yet ultimately hollow. But now, we move from metaphor to reality.

## How Evolution Devalues Human Life

Darwin didn't explain the origin of life—only proposed a model that removed its significance. This shift had consequences not only for society—but for Darwin himself. In his *Autobiography*, he wrote:

*I gradually came to disbelieve in Christianity as a
divine revelation... This disbelief crept over me at a
very slow rate, but at last it was complete.*[1]

What began as inquiry led to abandonment of faith—and
Darwin was not alone. His descent echoed through generations,
influencing many.

The same rebellion appears in Friedrich Nietzsche's words:
*"God is dead."* But history hasn't vindicated his worldview.
Nietzsche's philosophy has withered. God's word endures—
*unchanged and eternally relevant.*

## Darwin's Allies: Secular Humanism and Atheism

Darwinian evolution is more than a theory—it's the ideological
backbone of secular humanism and atheism. Both belief systems
rely heavily on evolution to reject divine creation. Without it, their
foundation would crumble.

At the heart of secular humanism lies a moral framework not
just independent of Scripture—but directly opposed to it. It
replaces God's authority with subjective ethics and human
autonomy.

This contrast isn't subtle—it's stark. The pursuit of godlessness
demands a redefinition of morality, purpose, and even reality itself,
leaving people unmoored—adrift in a sea of uncertainty, and
spiritual disillusionment.

## A Clash of Worldviews: Biblical Theism vs. Secular Humanism

This section highlights a striking contrast: *God's way*, as revealed in Scripture, versus *the world's way*, shaped by human reasoning and secular philosophy.

These worldviews aren't just different—they are *diametrically opposed*. They offer radically different answers to life's biggest questions: Who is God? What is truth? Why are we here?

Let's walk through ten key differences that reveal just how deeply these opposing worldviews diverge—starting with the most foundational: our view of God.

## Comparison 1: The True God vs. Man-Made Gods

*God's Way:* The Bible reveals one true God—Father, Son, and Holy Spirit.

*The World's Way:* Denies or distorts the biblical God in favor of atheism, agnosticism, false gods, or man-made belief systems.

*When belief in the true God is abandoned, societies don't remain neutral—they turn to substitutes. These may include secular ideologies, materialism, counterfeit religions, or spiritual distortions. But none can offer the truth, peace, or salvation found in the God of Scripture.*

## Comparison 2: Objective Truth vs. Relative Truth

*God's Way:* Truth is objective, eternal, and unchanging.

*The World's Way:* Truth is seen as relative—shaped by personal feelings, cultural trends, and individual moral preferences.

*Can a society truly thrive without a foundation of unchanging truth? Or does the erosion of that foundation lead to confusion, division, and moral collapse?*

## Comparison 3: Divine Creation vs. Accidental Evolution

*God's Way:* God created the universe with intention, purpose, and design. Human life is no accident. We're made in God's image.

*The World's Way:* Life is the result of random, unguided evolutionary processes—without purpose or divine direction.

*Belief in divine creation shapes how we view identity, dignity, and purpose. But what happens when that truth is replaced by the belief that we are merely cosmic accidents?*

## Comparison 4: Sinful Nature vs. Inherent Goodness

*God's Way:* Humanity is sinful and in need of redemption through Jesus Christ.

*The World's Way:* People are seen as basically good and not in need of a Savior.

*What does this belief say about our need for forgiveness and grace? Does confidence in human goodness bring redemption—or just the illusion of self-sufficiency?*

## Comparison 5: Reverence for Christ vs. Contempt for His Name

*God's Way:* Jesus is honored as Lord. Sin is confronted in love, without compromise.

*The World's Way:* Jesus' name is often treated with irreverence or used as profanity, while sin is normalized and righteousness ridiculed.

*What are the cultural outcomes of mocking what is holy and celebrating what is harmful?*

## Comparison 6: Moral Responsibility vs. Personal Autonomy

*God's Way:* Emphasizes accountability, self-control, and forgiveness through Christ.

*The World's Way:* Prioritizes personal autonomy—often at the expense of accountability—leading to indulgence and resentment rather than repentance.

*What happens when a society elevates personal freedom over moral responsibility?*

## Comparison 7: The Sanctity of Life vs. the Right to Choose

*God's Way:* All life—born and unborn—is sacred, created in God's image.

*The World's Way:* Elevates individual autonomy above the value of unborn life, framing abortion as a personal right.

*Can a society claim to value life while defending the right to end the lives of its most innocent and vulnerable? The hypocrisy is heartbreaking.*

## Comparison 8: God's Design for Family vs. Cultural Redefinition

*God's Way:* Marriage is between one man and one woman. Gender is intentional and unchanging—male and female, by God's design.

*The World's Way:* Redefines family, marriage, and gender based on personal identity, feelings, or cultural trends.

*What are the long-term consequences when society abandons the blueprint God gave for marriage, gender, and family?*

## Comparison 9: Sexual Purity vs. Sexual Exploitation

*God's Way:* Sex is sacred and reserved for marriage between one man and one woman.

*The World's Way:* Encourages sexual behavior outside God's design—whether through promiscuity, pornography, or same-sex relationships—degrading the dignity and sanctity of human sexuality.

*What happens when sexual boundaries are erased and replaced by personal desire?*

## Comparison 10: Eternal Focus vs. Worldly Focus

*God's Way:* Life is lived with eternity in mind—worshiping the Creator, growing spiritually, loving others, making disciples, and faithfully sharing the gospel of salvation. These eternal priorities take precedence over temporary pleasure.

*The World's Way:* Prioritizes the temporary—like personal success, material gain, or social and environmental causes. It places more emphasis on adoring creation itself—whether possessions, status, or even nature—while neglecting the Creator.

Caring for the environment and addressing worldly issues has value. But too often, they take center stage—overshadowing the ultimate concern: the salvation of souls. When temporary concerns become ultimate priorities, spiritual truth is sidelined, and eternal purpose is lost.

*What's the danger of focusing on the here-and-now—elevating the temporal and worshiping created things—while ignoring the eternal and the One who made it all?*

These ten comparisons reveal far more than surface-level disagreements. They expose a deep, spiritual divide—one that shapes how people believe, behave, and build their lives.

## The Battle Between Truth and Lies

How did humanity come to hold two worldviews so radically opposed—one grounded in biblical truth, and the other shaped by subjective morality? The Bible provides the clearest answer: we are caught in a timeless struggle between good and evil, a spiritual battle spanning every generation.

This battle begins in Genesis 3, when Adam and Eve disobeyed God—launching a war that continues to shape human history and the destiny of souls.

But this conflict didn't start with us. It began in heaven—with an ancient rebellion. Satan, driven by pride, sought to usurp God's throne. Isaiah records his bold declaration:

*I will make myself like the Most High. — Isaiah 14:14*

This arrogant desire to rival God became the blueprint for every rebellion that followed. It is the same spirit that fuels worldviews which reject divine authority in favor of human autonomy.

Consider the thread connecting secular humanism, atheism, and spiritual movements like New Age. Though distinct in form—from the supremacy of human reason, to the outright denial of God, to the belief in an inner power to create one's own reality—each ultimately seeks to dethrone God and enthrone the self.

The ancient rebellion set the stage for an enduring struggle—between obedience and defiance, flesh and spirit, God's will and human pride. These worldviews reveal a deeper spiritual reality: apart from God, humanity gravitates toward what is harmful and self-serving. Instead of pursuing good and truth, we chase desires that defy His design, leading to confusion, ruin, and death.

The spirit of this age blinds hearts and minds, replacing biblical truth with false ideologies. Many are swept into a self-defined morality that distorts justice, wounds individuals, fractures society, and draws us further from the abundant life God intends.

Jesus made it clear: there is no neutral ground:

*Whoever is not with me is against me, and whoever does not gather with me scatters. — Matthew 12:30*

Turning away from God is nothing new. Psalm 2:2 warns:

*The kings of the earth rise up and the rulers band*
*together against the Lord and against his Anointed...*

But this isn't merely a human rebellion—it's a spiritual battle.
Paul reminds us:

*For our struggle is not against flesh and blood, but*
*against the rulers, against the authorities, against the*
*powers of this dark world and against the spiritual*
*forces of evil in the heavenly realms. — Ephesians 6:12*

This rebellion continues to impact our world. Each must ask:
Will we be filled with the Holy Spirit—the Spirit of Truth—or
yield to the influence of the evil one, who rejects truth and
promotes deception?

The divide between biblical theism and secular humanism is
stark. It establishes a clear line of demarcation—one that either
strengthens or undermines faith.

Such profoundly opposite worldviews with conflicting moral
and spiritual foundations could hardly arise by chance. The idea
that these belief systems emerged without intentional spiritual
influence is not just unlikely; it borders on absurdity. This divide
reveals something deeper: the nature of God and humanity's
ongoing rebellion against Him.

Amid this cosmic conflict, Jesus declared His purpose:

*In fact, the reason I was born and came into the world is to testify to the truth. Everyone on the side of truth listens to me. — John 18:37*

Personally, this realization came after years of mental and spiritual wrestling—even in sleep, where I felt the weight of spiritual warfare pressing on my spirit.

I do not write from a distance. My heart grieves for loved ones who have walked away from faith. The pain of watching someone reject truth is sorrow I know well.

These experiences remind me that this battle is not theoretical—it's real and deeply personal. Once I saw humanity's defiance against God—including my own—I could not unsee it. This awakening was a turning point in my faith, reshaping how I understand God's standards, His purpose, and the reality of the spiritual war we all face.

Even more sobering is the growing confusion among believers. Increasingly, many professing Christians are embracing blended worldviews that mix truth with error, light with darkness, and wisdom with folly.

Tragically, many—even within the church—have become captive to ideologies that oppose God's heart.

*...and that they will come to their senses and escape from the trap of the devil, who has taken them captive to do his will. — 2 Timothy 2:26*

This raises a critical question: Will we continue embracing humanistic tendencies, or will we turn to the living God of the Bible for clarity and strength to overcome them?

In a world clouded by confusion, may we return to the true mirror—God's Word—where our calling, our hope, and our worth remain unshaken.

**If this resonates with you, consider pausing to pray:**

*Heavenly Father, in a world filled with so much confusion and conflicting ideas, we ask for Your truth to shine brightly. Help us to discern Your voice and Your Word amidst the noise. Humble our hearts to accept what You declare as truth, even when it challenges our own understanding or the prevailing culture. Fill us with Your Holy Spirit, the Spirit of Truth, so that we may live in obedience to You and faithfully testify to the good news of Jesus Christ, who came into the world to reveal Your perfect truth. Thank you for Your incredible love for us. Amen.*

**Finally, remember this:** The enemy has always tried to counterfeit the truth—offering evolution in place of creation, relativism instead of absolute truth, and lust in place of love.

But *God is real.*

*His truth still stands.*

*And Christ still saves.*

## The Role of Free Will

At the heart of the battle for truth is **free will**—the God-given ability to choose between good and evil, truth and lies. Scripture teaches that God endowed humanity with this gift, allowing love, repentance, and salvation to be genuine. Without free will, there would be no moral responsibility or meaningful relationship with God. Love must be freely given—not forced. We are not spiritual robots.

In contrast, the secular evolutionary worldview often minimizes—or outright denies—free will. If humans are merely products of chance and natural selection, then what we call "free will" is nothing more than an illusion—just the result of unguided biological processes. In that view, our behavior isn't guided by moral choice, but by instincts, genetics, and environment. Even our thoughts aren't truly chosen; they're seen as neurochemical patterns shaped by heredity and experience.

Through this lens, humans become victims of biology and circumstance. Moral accountability dissolves, truth becomes subjective, and justice becomes negotiable. Right and wrong lose their foundation—reduced to evolving opinions rather than enduring truths.

From a Darwinian perspective, we are no longer image-bearers of God with the power to choose truth and love. We are viewed instead as accidental byproducts of a purposeless process. Free will is dismissed. Moral reasoning is considered irrelevant—or even naive. Without divine design, there is no higher calling—only survival.

But this battle for truth is not just intellectual—it is spiritual. What's at stake is more than theory or opinion. It's about who we are, and who we are becoming. Are we morally responsible beings—redeemed by grace and made in God's image—or aimless creatures ruled by instinct and impulse?

The choice is clear: either embrace the true freedom God offers, or surrender to philosophies that deny our dignity and responsibility.

Our decisions matter. They shape our character, our culture, and our destiny.

True freedom—the freedom to love, to choose, to follow God—can only exists when we recognize our divine origin and moral accountability. In a world increasingly set against both, choosing God's truth is the most powerful act of freedom we can make.

## The Consequences of Relative Truth

Imagine a world where everyone creates their own version of what's right—where facts are fluid, and reality bends to personal preference. At first, this might sound freeing. But the cracks quickly appear.

Take math, for example. What if a student insisted that 2 + 2 equals 5 simply because it "felt right" to them? If their version of reality is as valid as the teacher's, then nothing adds up—literally. Without fixed truth, math collapses—and so does everything built on it.

The same is true for morality. When it becomes relative, society loses its ability to function with order, justice, and peace.

Or picture hiring a builder to construct your dream home. You hand over carefully drawn plans, but your builder decides to "interpret" them differently. Instead of a sleek modern rambler, you end up with a crooked, upside-down colonial. Who's right? In a world of relativism—*everyone is, and no one is.*

Now consider something even more serious: driving. If stop signs are optional and speed limits are mere suggestions, chaos erupts. When each person drives by their own rules, the road becomes a battleground. Lives are lost when shared understandings no longer hold.

The consequences of such a culture are clear: **confusion, disorder, and eventual collapse.**

But there's a better way.

God offers absolute truth—unchanging, trustworthy, and rooted in His character. The Bible is not a random rulebook; it's a blueprint for life, written by the One who designed it. In a world of shifting opinions and moral gray zones, God's truth brings **clarity, stability, and peace.**

Truth isn't invented—it's discovered. And the truth that comes from God doesn't just make more sense—it sets us free: free from confusion, free from fear, and free from the power of sin.

This is the true freedom found only in Christ. As Jesus Himself declared:

> *So if the Son sets you free, you will be free indeed.*
> *— John 8:36*

## America's Descent into Secularism

The 1960s marked a spiritual crossroads for the United States. Legal decisions that removed prayer and Bible reading from public schools didn't merely eliminate religious practices—they began dismantling the foundation of a biblically grounded worldview in education. Into that void came a new ideology: Darwinian evolution. Though presented as objective science, it often functions as a secular framework, reshaping how generations understand life, purpose, morality, and identity.

This was not a neutral exchange of ideas. It signaled a profound cultural shift with far-reaching consequences. The God-centered vision that once guided our nation was gradually replaced by secular humanism—a man-centered belief system built on the denial or distortion of divine truth.

To be clear, this is not a rejection of science itself. Many of history's greatest scientific minds were deeply committed Christians who saw no conflict between faith and reason. Figures like **Sir Isaac Newton, Johannes Kepler, Blaise Pascal**, and **Robert Boyle** were driven by a belief in a rational Creator whose universe could be studied and understood. Even today, many respected scientists across disciplines—from physics to biology—hold to a biblical worldview.

The issue is not with science, but with how it is sometimes misused to advance an atheistic or materialistic agenda that denies our divine origin.

While biblical Creation has been marginalized or mocked in public institutions, evolution is publicly funded, widely promoted, and largely protected from critical scrutiny. Those who question it are frequently dismissed, censored, or ridiculed.

The result? A society adrift—*cut loose from its moral anchor, tossed in a storm of confusion and division.*

## The Tragic Ripple Effect

As Darwinian evolution became the dominant lens through which students viewed the world, belief in absolute truth began to crumble. In its place rose a culture steeped in moral relativism and human-centered pride. This shift is evident in the rise of politicized science, confused sexual ethics, and the erosion of traditional moral foundations. We see capitalism giving way to socialism, and altruism replaced by growing narcissism.

This is the "fault line" of a nation on the verge of collapse—where moral confusion, spiritual rebellion, and cultural decay all converge.

**This worldview shift has not been without cost. We're witnessing a society unraveling—marked by skyrocketing divorce rates, gender confusion, abortion, school shootings, and sexual exploitation.**

**Add to this the rise in STD rates, widespread drug use, deepening depression, and a heartbreaking surge in teen suicides.**

According to the Centers for Disease Control and Prevention (CDC), the suicide rate among youth ages 10 to 24 has risen by over **100%** since the 1960s.[2] School shootings, once unthinkable, now occur with disturbing regularity.

In the 1950s and 60s—especially in rural areas—it wasn't unusual for students to bring rifles to school. Guns were stored in lockers or visible in pickup trucks parked outside.[3]

Yet violence was rare.

Clearly, something far deeper has changed. These aren't isolated issues—they are the inevitable consequences of a nation that has abandoned God's truth and embraced the lies of secularism.

When we teach children that they are nothing more than the product of evolution, mere animals shaped by random forces—it should not surprise us when they begin to act with recklessness, lawlessness, and a disregard for the value of human life.

The moral fabric of society isn't just fraying—it's being actively unraveled.

Behaviors once considered immoral or destructive are now increasingly normalized. What was once shameful is now celebrated. What was once condemned is now praised as virtuous.

Those who believe that affirming sinful behaviors under the banner of the "Woke" agenda is an act of love or compassion should consider the fruit it bears. As cultural approval of these behaviors has increased, so too have rates of depression, anxiety, anger, and even violence—especially among the very people this ideology claims to help. What is often labeled as "inclusive" or "affirming" may in fact be exacerbating brokenness and confusion.

I've been saddened—along with many others—by how often disagreement is equated with hatred. I don't hate anyone. In fact, I believe we're called to love all people—even if we don't affirm all behaviors. That distinction is critical. For example, I can deeply love someone who struggles with alcoholism, while still believing that getting drunk is harmful and wrong.

Jesus never compromised truth to show love, and He never withheld love in the name of truth. He did both—perfectly. That's the example we're called to follow.

Is America spiraling into a moral tailspin?

EVOLUTION'S RESPONSE TO GOD'S LOVE

Sadly, a defiant voice resonates within youth culture, rejecting God's tender mercy and infinite wisdom amid society's moral crisis.

The rise of secular humanism—driven largely by evolutionary teaching—has been a dangerous experiment, endorsed by the state and funded by taxpayers. What makes it even more egregious is that it continues while biblical alternatives are silenced.

**Charles Darwin's ideas about evolution have fueled the most pernicious indoctrination campaign ever perpetrated on humanity.**

This may sound extreme—but it's not hyperbole. The cultural fallout speaks for itself.

For generations, the theory of evolution has been strategically positioned as "settled science," while any mention of divine design has been censored, mocked, or outright banned in academic settings. That's not education—it's indoctrination. When only one narrative is allowed, and dissent is punished, we are no longer fostering critical thinking—we are grooming ideological conformity.

Isaiah's warning rings loud and true:

*Woe to those who call evil good and good evil*
*— Isaiah 5:20a*

This ancient cry speaks directly to our time, as we witness a growing reversal of good and evil in today's culture.

## Signs of the Cultural War

In Minnesota, this ideological shift is playing out in a particularly striking way. During the 2024 Christmas season, a satanic-themed display was installed in the State Capitol—a symbol of the growing influence of secularism and the rejection of God. The display featured a phoenix with an inverted pentagram above it, a well-known symbol of Satanism.

Over the past five years, several other states have seen similar displays in public spaces:

- **Illinois (2022):** A Satanic nativity and serpent-themed display at the State Capitol.

- **Iowa (2023):** The Satanic Temple placed a Baphomet statue at the State Capitol.

- **Wisconsin (2023):** A holiday tree adorned with pentagrams at a public festival.

- **New Hampshire (2024):** A Satanic nativity scene outside the Statehouse.

- **Kansas (early 2025):** The Satanic Grotto's "Black Mass" at the State Capitol.

These displays are part of a broader cultural and spiritual battle in which symbols of rebellion are not just tolerated but often protected and promoted—while Christian symbols are increasingly pushed aside. Jesus Himself forewarned His followers of such opposition, stating:

> *If the world hates you, keep in mind that it hated me first. — John 15:18*

This reality underscores that the conflict is not merely cultural, but deeply spiritual, reflecting the world's historical rejection of Christ Himself.

Though more people are waking up to the connection between secular ideologies and cultural decline, the path we've chosen has led to profound, often irreversible effects. The continued promotion of biological evolution in schools remains *ground zero* in the culture war. This foundational indoctrination causes many young people to question Scripture, embrace postmodernism, and sever themselves from God's truth.

Humanity has gone to great lengths to silence God from their consciences, but God, in His mercy, continues to extend freedom and redemption. The prophet Isaiah rightly captures the condition of our nation:

*Woe to those who go to great depths to hide their plans*
*from the LORD, who do their work in darkness and*
*think, "Who sees us? Who will know? — Isaiah 29:15*

When Christians try to merge the Bible with evolutionary theory, confusion follows. A faith built on the shifting sands of relative truth cannot stand. The line is becoming clearer between believers grounded in God's Word and those compromised by secular thought.

Darwin's theory has become the *philosophical linchpin* of atheism and secular humanism—exalting man while pushing God to the margins. Remove evolution, and the entire secular house of cards begins to collapse. Without evolution, atheism loses its last intellectual excuse for rejecting God. As Sir Arthur Keith, a respected British anthropologist, once admitted:

*Evolution is unproved and unprovable. We believe it*
*only because the only alternative is special creation,*
*and that is unthinkable.*[4]

Our money may still say "In God We Trust," but in practice, our actions increasingly shout, **"In Man We Trust."**

Evolution is the cornerstone upon which secularism and humanism rest—exalting man and disregarding God. This rejection of special creation is not merely theoretical—it's a deliberate and spiritual rebellion, seeking to disconnect humanity from its divine origin.

As we continue down this path, the effects are increasingly undeniable: a society that denies God's truth will ultimately descend into moral and spiritual chaos.

## The Roots of Evil

Darwin's theory of evolution has played a central role in some of the most horrific ideologies in history. The phrase *"ideas have consequences"* rings chillingly true here.

**Adolf Hitler**, deeply influenced by Darwinian thought, believed he was advancing the "master race" by eliminating the weak. In *Mein Kampf,* he wrote:

> *The stronger must dominate and not mate with the weaker, which would signify the sacrifice of its own higher nature. Only the born weakling can look upon this principle as cruel, and if he does so, it is merely because he is of a feebler nature and narrower mind; for if such a law did not direct the process of evolution, then the higher development of organic life would not be conceivable at all.*[5]

Hitler used evolution to justify genocide, claiming he was simply accelerating nature's process to produce a stronger race.

But he was not alone. Dictators like **Stalin, Mao Zedong, and Pol Pot** were also shaped by Darwinian thinking, combined with Marxist ideology. Stalin, after abandoning his seminary training upon reading Darwin, adopted an atheistic, materialist worldview that he used to rationalize mass murder.[6] These regimes were responsible for tens of millions of deaths—the worst atrocities in recorded history.

Dr. **D. James Kennedy** observed:

> *It was Darwin's theory—carried to its logical conclusion—that led to the death of some 11 million people at the hands of German Nazis. Hitler was a devout evolutionist. He instructed his troops in evolution and had them provided with books by Darwin and Friedrich Nietzsche.[7]*

The eugenics movement, which led to forced sterilizations and racial genocide in the 20th century, also drew heavily from Darwin's concept of "natural selection." Even the full title of his book reveals this underlying belief:

> *On the Origin of Species by Means of Natural Selection, or the Preservation of Favoured Races in the Struggle for Life.*

This idea of "favored races" gave pseudo-scientific legitimacy to horrific acts of dehumanization. It exposed a dark undercurrent of Darwinian thought that led to immense suffering and death.

Dr. Kennedy went further:

> *These communist leaders and others killed more people than all those killed in all religious wars combined... Stalin, Mao, Pol Pot, and all the rest are the greatest mass murderers of all time—and all compliments of evolution.*[8]

Former atheist and evolutionist **Tom DeRosa** traced his own path to disbelief directly to Darwin's work:

> *I read Origin shortly after I completed my high school studies at a Roman Catholic seminary. I had been on my way to entering the priesthood, but Origin made me question my faith. It raised doubts that led me to become an atheist and denounce God while attending college in New York City. My return to the city and my long subway ride brought to mind how desperately I once wanted to be 'free' in thought and in action and how Darwin offered me an escape from God—one I eagerly took.*[9]

In sum, Darwin's theory has not only shaped science—it has fueled some of the darkest ideologies in human history, from eugenics to totalitarian genocide. While not everyone who accepts evolution embraces these extremes, the philosophical consequences of Darwinian thinking remain deeply troubling.

And yet, these connections are rarely acknowledged. Public schools may teach about the Holocaust or communist oppression—but often ignore the ideological foundations that helped justify them. The influence of Darwinism on movements like eugenics, communism, and mass murder deserves far greater scrutiny than it currently receives.

## The Dark Legacy: Eugenics and Abortion

Few realize that the roots of modern abortion advocacy are deeply intertwined with the eugenics movement—a philosophy rooted in Darwinian thinking. Charles Darwin's cousin, **Francis Galton,** pioneered eugenics, promoting the sterilization of individuals labeled "inferior." This ideology shaped early 20th-century social policies and laid a troubling foundation for modern abortion practices.

**Margaret Sanger**, founder of Planned Parenthood, was a vocal supporter of eugenics and espoused openly racist views—particularly targeting African American communities. According to Heritage Foundation:

> *One of Planned Parenthood's largest abortion affiliates has finally disavowed Planned Parenthood's founder*

*for "her racist legacy" and her "connections to the eugenics movement." However, this symbolic bowing to the far left's "cancel culture" doesn't change the fact that the organization is still influenced by her inhumane beliefs.[10]*

Planned Parenthood of Greater New York (PPGNY) removed Margaret Sanger's name from its Manhattan clinic after decades of choosing to overlook the organization's white supremacist roots.

Sanger made numerous deeply disturbing statements. In one instance she wrote:

*We don't want the word to go out that we want to exterminate the Negro population.[11]*

She openly favored the forced sterilization of those she considered "unfit," and even gave a speech to the Ku Klux Klan.

*The most merciful thing that the large family does to one of its infant members is to kill it.[12] (Notably, she was referring to a child who had already been born— an idea akin to infanticide, which has unfortunately resurfaced in some modern political debates.)*

Tragically, her legacy lives on in the millions of abortions performed each year, disproportionately impacting low-income and minority communities.

Her beliefs in population control rested on the idea that certain groups were less worthy of life. This mindset helped propel policies that targeted vulnerable populations. Even today, the ongoing push for abortion access, particularly in areas with significant Black and Hispanic populations, reflects the lingering influence of these harmful ideologies.

Echoes of eugenics still resonate in modern abortion debates—with some viewing it as a tool for eliminating those deemed "undesirable." This dark legacy continues to undermine the sanctity and value of human life, particularly among the most vulnerable—perpetuating a dangerous historical pattern.

## The Urgency of Confronting False Ideologies

In confronting these deeply rooted ideologies, we're not just engaging in a cultural debate—we're entering a spiritual battle. Scripture is clear about how we are to engage in this conflict:

> *The weapons we fight with are not the weapons of the world. On the contrary, they have divine power to demolish strongholds. We demolish arguments and every pretension that sets itself up against the knowledge of God, and we take captive every thought to make it obedient to Christ. — 2 Corinthians 10:4–5*

As Christians, we are called not only to live out our faith in love but also to engage the world with truth and discernment. This passage reminds us that it's not just okay—it's necessary—to lovingly challenge ideas opposed to God's truth. Not by force, but through persuasion, prayer, and the power of the Spirit. The real battle isn't against people but against deception and falsehood.

*Have nothing to do with the fruitless deeds of darkness, but rather expose them. — Ephesians 5:11*

These verses don't give us license to be harsh, condescending, or proud—but they do call us to speak with clarity and courage, combining truth and love. Remaining silent or affirming falsehoods to avoid conflict isn't loving—it's actually harmful.

As we've seen, the impact of Darwinian evolution is far from abstract or theoretical. It has fueled destructive ideologies—including eugenics, abortion, and genocide—and has helped erode the moral fabric of our society. The rejection of divine truth has had devastating consequences, affecting not only our beliefs but also our actions and the very way we understand what it means to be human.

Evolutionary theory has become the cornerstone of secularism, exalting man while diminishing the Creator. This philosophical shift has led to moral decay, cultural confusion, and a society adrift without a moral compass.

Where will this path lead? This is more than an intellectual debate—it's a spiritual crossroads. Will you choose truth, or cling to a narrative that cannot even explain the reality we live in?

If these sections have stirred your heart and challenged your thinking, we invite you to explore Stumbling Block #16: *The Bible vs. Evolution: The Ultimate Collision*—and discover how to begin a life-changing relationship with Jesus Christ.

**The choice is clear:** continue down the road of secular humanism and its devastating legacy—or return to the unshakable foundation of God's Word, the only truth that leads to life.

This is no longer merely a matter of information—it is a matter of eternity.

Will you choose truth, and the life God offers, or embrace the lie that leads to destruction?

The consequences are real—but so is the invitation of Jesus:

*Come to the truth. Be set free. Live the life you were created for.*

It's time to choose wisely—and to choose truth. Let the truth set you free. This is the call. *Will you answer?*

# Evolution in Schools

# Is There a Hidden Bias?

*In classrooms where questions are quickly denied,*
*The truth of our origins is swept aside.*
*For evolution's tale, no room to debate,*
*Silences beliefs we should not forsake.*

The poem above captures a growing concern: that modern education often discourages open inquiry when it comes to life's origins. To illustrate this, let's begin with a story.

## The School of Glass

There was once a grand school in a city of stone, known far and wide as the *School of Glass*. Its founders had built it upon a foundation of light—sunlight poured in through stained-glass windows that told stories of truth, beauty, and a Creator who made all things with purpose.

Each classroom was a place of wonder. Students were encouraged to ask questions, wrestle with big ideas, and seek truth with open hearts. Teachers guided them not to conformity, but to clarity—to see the fingerprints of design in both nature and soul.

But as time passed, the windows began to crack.

A new Headmaster was appointed. He called himself a "modernizer." Under his leadership, the stained glass was removed and replaced with polished, tinted panes—mirrors from the *Institute of Pure Science*. These mirrors looked outwardly sleek, but they allowed only one type of light to enter—the light of material explanation.

"No more fables," the Headmaster said. "Only facts."

A new creed was etched on the front gate:

*"From chaos we came. By chance we remain. Purpose is a myth."*

At first, students questioned it. But soon, curiosity was replaced by compliance. Teachers who refused the creed were quietly dismissed. Books that told of a Designer were shelved, then banned. When students asked about other explanations for life's origin, they were told: "That's not science. That's religion. And religion has no place here."

Over time, something shifted. The wonder faded. The questions stopped. Eyes dulled. Hearts grew heavy. In the absence of purpose, discipline waned, kindness waned, even truth itself became negotiable. The school was full of facts—but empty of light.

One girl, brave and restless, whispered to a friend: "Isn't there more than this?"

He looked toward the old storeroom, where the stained glass had been locked away. "There used to be," he said. "But we're not allowed to look."

In a classroom once illuminated by the light of truth and wonder, we now find ourselves in the shadow of a single, narrow view—a view that refuses to let other perspectives shine through.

Though fictional, the School of Glass reflects a growing reality in American education. Classrooms once devoted to discovery and diverse thought now often limit discussion to a single, state-approved narrative.

When the state promotes a secular worldview in this way, it crosses a critical line—raising not only educational issues but serious constitutional questions.

## When Purpose Was Removed from Education

For over 50 years, the U.S. government has been advancing the ideologically driven philosophy of Secular Humanism by presenting evolution in public schools as the exclusive explanation for the origin of life.[1]

This approach, often framed as purely scientific, promotes a specific worldview—raising serious constitutional concerns.

The Constitution's Establishment Clause forbids the government from establishing a religion or promoting one belief system over others.

The key word here is *establishing*—meaning the government is prohibited from creating, endorsing, or favoring any particular religion or ideology, whether theistic or non-theistic.

By promoting Secular Humanism through the exclusive teaching of evolution in public schools, the government is advancing a secular ideology that effectively functions as a religion. This is a clear violation of the Constitution.

## How Can We Be Sure That Secular Humanism Operates as a Religion?

To understand why, it's essential to first consider what generally defines a religious belief system. Religious beliefs, whether theistic or non-theistic, typically share two core characteristics:

1. **They establish a comprehensive worldview that defines moral values and guides behavior.**

2. **They provide foundational narratives concerning humanity's origins and ultimate purpose (or lack thereof).**

By contrast, disciplines such as physics, chemistry, and mathematics are grounded in empirical observation and do not shape moral values in the same way.

All religions generally fall into two broad categories: *theistic* (those that believe in a god or gods) and *non-theistic* (those that do not). Secular Humanism can certainly be classified as a non-theistic religion—it denies the existence of a supernatural deity and asserts that nature is all there is. This stands in stark contrast to theistic religions, which affirm the existence of a spiritual reality beyond the physical world.

In both cases, whether belief in God or the belief in the absence of God, a measure of faith is required to address ultimate questions that cannot be answered solely through empirical observation. Therefore, the promotion of evolution through the lens of Secular Humanism is ultimately rooted in faith, just as much as any theistic religion.

Should any doubts remain, these candid admissions from Humanist leaders make it clear:

– **Charles Francis Potter**, a signer of the *Humanist Manifesto* and author of *Humanism: A New Religion*, stated:

> *Education is the most powerful ally of Humanism, and every American public school is a school of Humanism. What can the theistic Sunday Schools, meeting for an*

*hour once a week, and teaching only a fraction of the children, do to stem the tide of a five-day program of humanistic teaching?*[2]

– **John J. Dunphy**, writing in *The Humanist Magazine*, echoed a similar view:

*I am convinced that the battle for humankind's future must be waged and won in the public school classroom by teachers that correctly perceive their role as proselytizers of a new faith: a religion of humanity... The classroom must and will become an arena of conflict between the old and new—the rotting corpse of Christianity, together with all its adjacent evils and misery, and the new faith of humanism...*[3]

These striking statements make it clear: proponents of Secular Humanism fully understand the role that public education plays in advancing their secular worldview.

The classroom becomes not merely a place for scientific instruction, but a platform for instilling a particular belief system.

This underscores how the government's promotion of evolution in public schools amounts to endorsing a specific worldview—religious in function, non-theistic in nature—and thus violates the constitutional principle of religious neutrality.

## Legal Recognition of Secular Humanism as a Religion

It's not only Humanists who view their movement as a religion—some U.S. courts have affirmed this status as well. In the landmark 1961 case *Torcaso v. Watkins*, the Supreme Court explicitly stated:

> *Among religions in this country which do not teach what would generally be considered a belief in the existence of God are Buddhism, Taoism, Ethical Culture, Secular Humanism, and others.*[4]

This Supreme Court ruling is pivotal, as it clearly includes Secular Humanism among recognized religions, specifically noting its non-theistic nature. This directly supports the understanding that, for constitutional purposes, a belief system does not require a supernatural deity to be considered a religion.

Additionally, in *Reed v. Van Hoven*, a federal court emphasized that public schools must "carefully avoid any program of indoctrination in ultimate values" between theistic and humanistic religions.[5]

These legal precedents confirm that Secular Humanism is not only embraced as a religion by its advocates but also acknowledged in legal rulings as a belief system subject to First Amendment protections and constraints.

By placing it alongside other worldviews, the judiciary underscores the critical importance of religious impartiality in public education. This principle cannot be overstated, as the failure to uphold such neutrality over the past five decades has led to tragic consequences.

As this secular worldview has gained prominence in schools—particularly through the exclusive teaching of evolution—we've witnessed skyrocketing rates of divorce, teen suicide, drug abuse, STDs, depression, and a widespread erosion of moral clarity.

Perhaps most tragically, this erosion of moral clarity extends even to our understanding of human dignity itself.

Today, an estimated 49.6 million people live in modern slavery—a pervasive, often hidden reality that encompasses human trafficking, forced labor, and forced marriage—more than at any other time in recorded history.[6]

This shocking figure underscores the profound consequences of a worldview that often removes purpose, accountability, and spiritual wisdom from the heart of education.

The promotion of Secular Humanism in this context violates the Establishment Clause and undermines the foundational values that once anchored American society.

## When Secularism Took Over the Classroom

In the 1960s, a series of landmark Supreme Court decisions fundamentally reshaped the landscape of public education in the United States. In 1962, the Court ruled against state-sponsored prayer in schools. In 1963, it banned Bible reading in classrooms.[7]

The most pivotal moment came in 1968 with *Epperson v. Arkansas*, which struck down a state law prohibiting the teaching of evolution.[8]

This decision effectively opened the door to a taxpayer-funded campaign to promote Secular Humanism under the banner of teaching Darwinian evolution.

In 1987, *Edwards v. Aguillard* pushed the door open further, striking down a law that required creation science to be taught alongside evolution.[9] With that ruling, evolution was cemented as the only legally sanctioned explanation of human origins in public schools.

Since these decisions, American society has experienced a steady and noticeable secular shift—a shift largely attributed to evolution's unchallenged dominance in classrooms. Consider several notable examples:

- Religious symbols and expressions are increasingly excluded from public spaces and schools—Ten Commandments displays removed, nativity scenes banned, and Christmas references (including "Christmas break" and sacred carols) secularized or censored.

- Graduation ceremonies often exclude prayer or any mention of God.

- Legal challenges target traditional religious phrases and symbols, such as "God" in the Pledge of Allegiance and the national motto, "In God We Trust."

- Student religious clubs face resistance despite legal protections.

- Textbooks have been revised to downplay or erase the role of faith in America's founding and history.

- The traditional calendar designations "BC" (Before Christ) and "AD" (Anno Domini—In the year of our Lord) are increasingly being replaced with "BCE" (Before Common Era) and "CE" (Common Era) in academic and educational materials to eliminate references to Christ.

- Voluntary personal expressions of faith—like student-led prayer, wearing crosses, or referencing Scripture in homework—are often discouraged or penalized.

- Curriculum in sex education and social studies increasingly reflects Secular Humanist values— promoting moral relativism, redefining marriage, gender and family structures, while marginalizing traditional religious beliefs.

What we are witnessing is not religious neutrality—it is the systematic exclusion of theistic faith, particularly Christianity, from the public square. This is most apparent in classrooms, where secular ideology is not just presented as fact but frequently enforced as the sole valid way to interpret the world.

This shift is not without consequence. It carries profound implications for the hearts and minds of current and future generations—shaping their moral compass, sense of purpose, and understanding of truth.

## Religious Freedom Under Fire

The ongoing legal struggle between secularism and religious freedom underscores the increasingly fragile state of constitutional impartiality.

In *Town of Greece v. Galloway* (2014), the Supreme Court ruled that opening government meetings with prayer did not violate the Establishment Clause—demonstrating how theistic religious expression is still being defended in some public settings.[10]

Yet in *American Atheists, Inc. v. United States*, and other cases, courts have entertained challenges to religious symbols on public property, even as secular ideologies continue to gain ground.[11]

In 2019, a federal appeals court struck down a California public school course that critics argued presented religious content through a Secular Humanist lens—raising serious concerns that secularism itself was being elevated as a state-endorsed belief system.

This case involved a mandatory world religions course, where critics claimed it portrayed Christianity in a negative light while favoring secular perspectives.[12]

Despite occasional victories for religious freedom, the broader trend remains clear: secularism continues to expand its reach in public institutions, often under the guise of upholding the *separation of church and state.*

## The Unintended Consequences of "Separation of Church and State"

Thomas Jefferson's phrase *"separation of church and state"* was never meant to remove theistic faith from public life.

Rather, it was intended to protect religious liberty by preventing the government from interfering in religious matters or establishing a national church.

Unfortunately, in recent decades, secularists have reinterpreted Jefferson's words to justify the exclusion of theistic beliefs—particularly Christianity—from public institutions, all in the name of "separation." This misuse is not only misleading—it's a contradiction of Jefferson's original intent.

Jefferson coined the phrase in an 1802 letter to the Danbury Baptists, reassuring them that the federal government would not interfere with their religious practices.[13]

Though the phrase *"separation of church and state"* appears nowhere in the Constitution, it is often confused with the First Amendment's Establishment Clause.

In fact, Jefferson's metaphor of a *"wall of separation"* was about limiting government power over religion, not erasing religion from the public square.

Modern secularists have inverted the concept. Instead of using the metaphor to restrain government overreach into religion, they've weaponized it to restrict religion from public life—especially in schools, courts, and civic institutions.

The result is a de facto endorsement of non-theistic worldviews like Secular Humanism, while theistic perspectives are increasingly marginalized.

This is not neutrality—it is preferential treatment based on a distorted interpretation of *"separation of church and state."*

The Founding Fathers' understanding of religious liberty was deeply shaped by Europe's legacy of state-sponsored religion. In England, the Church of England marginalized dissenters.

In Catholic nations like France and Spain, non-Catholics faced legal and social persecution. Many early American settlers—such as the Puritans and Pilgrims—fled to the New World seeking freedom to worship outside of state control.

These experiences led the Founders to include the Establishment Clause—not to purge public life of all religion, but to prevent the state from mandating one religious view over others.

Today, however, public schools do just that by promoting a worldview rooted in Darwinian evolution—one that aligns with Secular Humanism and systematically excludes any mention of divine design or purpose.

This is precisely the kind of ideological imposition the Founders sought to avoid. By endorsing a non-theistic belief system in the classroom while excluding theistic perspectives, the state is no longer neutral.

It is, in effect, promoting a secular religion—an act the Constitution was written to prevent. The consequences of this distortion cannot be overstated: not only does it erode genuine religious freedom, it also undermines the pluralistic and moral foundations on which the nation was founded.

Beyond legal concerns, the exclusive teaching of evolution has deeper implications—shaping not just minds, but worldviews.

## Evolution: The Hidden Faith

For many who embrace a secular humanist worldview, the theory of evolution functions as the cornerstone, much like Scripture serves as the foundation of Christianity.

In public education, evolution is often presented as "settled science" or an unquestionable truth.

Belief in evolution—particularly as an unguided, godless process—requires a measure of philosophical faith. It goes beyond empirical science and enters the realm of worldview, deliberately excluding the possibility of a Creator.

While science is grounded in observation and repeatable experimentation, the theory of evolution—particularly in its application to life's ultimate origins—often extends beyond empirical inquiry.

It is presented not only as a biological explanation, but as a comprehensive narrative for understanding life's origin, meaning, and morality. In this way, evolution functions not merely as a scientific theory, but as a non-theistic belief system—one that dismisses divine purpose and moral accountability.

Despite this, Secular Humanism enjoys what amounts to *"constitutional immunity."* It has largely avoided the scrutiny typically applied under the Establishment Clause.

Rather than maintaining neutrality, public schools have increasingly become platforms for advancing a non-theistic philosophical framework. In classrooms across the country, Darwin's theory is often taught with dogmatic fervor, while any critique—no matter how scientifically grounded—is treated as intellectual heresy.

This reveals a critical blind spot: many parents and educators do not realize that evolution, though labeled as factual science, effectively serves as a belief system—one that directly contradicts the deeply held convictions of countless families.

The paradox is stark: schools claim neutrality by banning theistic perspectives, yet simultaneously promote a secular worldview rooted in its own metaphysical assumptions.

## The Double Standard of Tolerating Evolution

Imagine the public outcry from secularists, the mainstream media, and other critics if schools were required to give equal time to Intelligent Design and biblical Creationism alongside evolution—ideas that directly challenge the foundation of the secular worldview taught in many classrooms today.

Many Christian parents mistakenly believe their children are simply being offered one perspective among many. However, evolution is not just another viewpoint presented in class—it's almost always the only perspective taught, with no room for alternatives.

Meanwhile, most churches avoid addressing the topic altogether, leaving parents ill-equipped to guide their children through this complex and deeply important debate.

In reality, public schools overwhelmingly promote evolution as the sole legitimate explanation for the origin of life, often misrepresenting or oversimplifying the debate.

While many parents would be alarmed if schools promoted controversial ideologies without balance, they often accept the unchallenged teaching of evolution—even when it conflicts with their core beliefs.

Why should secular worldviews not be held to the same level of scrutiny in public education as any other belief system—especially when such promotion may violate the Establishment Clause of the Constitution?

We are not calling for the removal of evolution from public education, nor are we seeking to impose religious indoctrination in science classrooms.

What we are advocating for is balanced education—an environment where students are encouraged to think critically by being exposed to more than one perspective on life's origins.

This includes allowing scientifically grounded alternatives, such as Intelligent Design, to be presented objectively and respectfully. It also means ensuring that textbooks use neutral and accurate language and do not portray materialist explanations as the only valid worldview.

True academic freedom fosters inquiry, encourages debate, and equips students to evaluate diverse ideas—not merely accept a single narrative by default. A balanced approach does not undermine science—it strengthens education.

## Restoring Balance in the Classroom

Parents who value both religious freedom and intellectual honesty in government-run schools should take an active role in defending constitutional principles and ensuring balance in the classroom:

- Why is intellectual freedom so restricted when it comes to discussing Darwin's theory?

- Why are students who question evolution often discouraged or silenced?

- Is it possible to remain unbiased in the culture war between creationism and evolution?

- If impartiality isn't possible, why not present both perspectives fairly and objectively, allowing students to decide for themselves?

- Why is evolution presented as an unquestionable fact rather than a theory open to critique?

- How can we justify teaching evolution as scientific fact when there are many unanswered questions and competing theories?

- Why are scientifically grounded alternatives, such as Intelligent Design or Creationism, excluded from meaningful classroom discussion?

- Are students being given the opportunity to critically analyze the evidence both for and against evolution?

- Does the exclusive teaching of evolution uphold the constitutional call for ideological neutrality in the classroom?

- What safeguards are in place to ensure that religious and philosophical impartiality is maintained in the classroom?

Raising these questions is the first step toward restoring fairness, transparency, and true educational freedom in our schools.

## How Parents Can Respond

Beyond simply asking these questions, parents should take action by organizing community meetings, speaking at school board meetings, and joining advocacy groups focused on educational reform and constitutional rights.

This proactive approach helps ensure that the conversation remains open and that educators are held accountable.

America's Founding Fathers warned against the dangers of state-sponsored religion. Today, we must recognize that the promotion of Secular Humanism in public schools reflects the very oppression they sought to avoid. By advocating for a balanced education that includes multiple perspectives, we ensure future generations can think critically, make informed decisions, and respect diverse viewpoints.

Upholding religious neutrality, intellectual freedom, and freedom of speech and religion in the classroom is not merely about protecting personal beliefs—it's about safeguarding the constitutional values of liberty, equality, and the fundamental right of each citizen to shape their own worldview.

We must insist that both sides of the debate on origins be presented fairly and honestly, giving students the freedom to explore and decide for themselves. Let's stand together for balanced education—and preserve the integrity of our educational system for future generations.

## Before You Turn the Page...

The clash over ultimate truth in our schools is not just political or academic—it's spiritual at its core. At the heart of this debate lies a deeper question: *What is the source of truth?*

If you've read this far and feel the weight of what's at stake—truth, eternity, and the next generation—keep reading. In Stumbling Block #16: *The Bible vs. Evolution: The Ultimate Collision,* **we'll explore the real story behind our origins and the lasting hope found in a life-changing relationship with Jesus Christ.**

*Don't stop at asking questions—keep seeking. What you find may change everything.*

# The Bible vs. Evolution

# The Ultimate Collision

*"In the beginning, God created the heavens and the earth."*

**— Genesis 1:1**

*Could God, in love, design a path*

*Where death and decay precede the wrath?*

*Evolution's claim, so cruel and long,*

*Seems to clash with where our faith belongs.*

## Death Before Sin? A Theological Dilemma

Could God have used evolution to create life?

This view—called *Theistic Evolution*—claims that God directed evolution over millions of years, weaving divine intention into mechanisms like mutations and natural selection.

At first glance, it might seem like a noble attempt to reconcile faith and science.

But it introduces a theological contradiction that strikes at the heart of the gospel—and the character of God.

Evolution teaches that death, disease, and extinction were present long before humans ever walked the earth. But Scripture tells a very different story:

> *Therefore, just as sin entered the world through one man, and death through sin, and in this way death came to all people, because all sinned. — Romans 5:12*

According to the Bible, death wasn't part of God's original creation. It was the consequence of human sin. This truth is foundational to the Christian faith:

**Sin brought death.**

**Death brought the grave.**

**But Christ conquered both.**

If death existed long before sin, then it wasn't a punishment at all—it was just part of how God made the world. And if that's true, we're forced to accept a troubling chain of implications:

- The curse came **before** original sin
- Judgment came **before** wrongdoing
- The cross solved a problem that existed **before** human rebellion.

## What Kind of God Would Choose Evolution?

Evolution is driven by death, extinction, and suffering. Could a loving, omnipotent God really use such a brutal process to create life? What kind of Creator would call that love?

If God is truly good, wouldn't He create through a process that reflects His nature—one marked by peace, order, and purpose?

A merciful God would not introduce pain, decay, and death *unless* it was to satisfy justice—not as part of His original design.

Anything else calls His goodness into question. If evolution—with all its violence and waste—were God's chosen tool, it raises unsettling doubts about His justice, His mercy, and even His worthiness of worship.

Even more troubling: how could God look upon a world built on death and declare it "very good"?

> *God saw all that He had made, and it was very good.* — *Genesis 1:31*

That statement only makes sense if the world was truly free from suffering—a place of peace, life, and harmony.

> *Nevertheless, death reigned from the time of Adam…* — *Romans 5:14*

Theistic evolution unravels this truth. If death isn't the result of sin, then the cross loses its meaning. If death is natural, Christ's resurrection is unnecessary. And if Genesis is unreliable, *where does biblical authority begin?*

The entire arc of redemption—from the fall to the cross to the resurrection—rests on one foundational truth:

*Death entered the world because of sin.*

## A Perfect Garden Built on Death?

Imagine the irony: Adam and Eve stroll through paradise—yet beneath their feet lies a graveyard, evidence of eons of death, extinction, and decay.

## Compromising Genesis Undermines the Entire Bible

Theistic evolution isn't a minor adjustment. It strikes at the very foundation of Scripture. When the clear teaching of Genesis is reinterpreted—or dismissed—the authority of the entire Bible begins to crumble.

If the origin of sin, death, and creation are open to debate, then no doctrine is safe. This erosion hasn't helped the Church reach modern culture—it's opened the floodgates to compromise.

A watered-down faith that conforms to secular theories:

- Dilutes the gospel
- Confuses believers
- Undermines confidence in God's Word

For over fifty years, the results have been tragic:

- Society drifting from biblical truth
- Churches paralyzed by cultural pressure
- A generation walking away from Scripture

These outcomes aren't random. They are the direct result of replacing biblical foundations with man-made theories.

## Why the Bible and Evolution Are Incompatible

The differences between the biblical creation account and evolutionary theory aren't just philosophical—they are theological, scientific, and foundational. Together, they form a deep, irreconcilable divide on several key issues:

- **The Creation Timeline:** The Bible teaches that humans and animals were created on Day Six, while evolution claims they arose from non-living matter over millions of years.

- **Order of Creation:** Genesis places plants before fish, fish before insects, and birds before reptiles—directly contradicting the evolutionary timeline.

- **Nature of the Event:** Scripture presents creation as a literal, historical act completed in six days—not a metaphor stretched across eons.

- **Words of Jesus:** Christ Himself affirmed the reality of creation and Noah's Flood (John 17:24; Mark 10:6; Matthew 24:37–39).

- **Human Uniqueness:** The Bible teaches that humans were made to rule over animals—not as evolved cousins, but as God's image-bearers.

- **Nature of God:** Evolution describes a process driven by randomness, death, and extinction. That doesn't reflect the nature of a loving and intentional Creator.

In the end, macroevolution discredits the Bible, and the Bible decisively discredits macroevolution. They cannot both be true.

If evolution is true, Scripture is wrong—especially in its foundational claims. But if Scripture is true, evolution collapses as a sufficient explanation for life.

## Clarity Comes from Trusting God's Word

When we start with Scripture, confusion clears. The creation timeline, the nature of life and death, and the uniqueness of humanity all point unmistakably to divine design—not to blind, random processes.

As followers of Christ, we are not called to reinterpret the Bible to fit secular theories. We are called to *uphold the truth* of God's Word—even when it's unpopular.

## Cultural Compromise: Eroding Biblical Doctrine

Some denominations have drifted so far into cultural accommodation that they now embrace unbiblical definitions of marriage, gender, and sexuality.

What God clearly established in Genesis—marriage as the union of one man and one woman, and humanity created as male and female—is now being questioned, redefined, or outright denied within parts of the Church.

This is not simply a matter of interpretation,

It is a surrender of biblical authority.

When Church leaders allow secular ideologies to shape doctrine, they lose moral clarity. The result is a Church that's spiritually adrift—confused, vulnerable, and unanchored from truth.

The moment we allow culture to dictate theology, we begin reshaping the Bible in our own image, rather than being conformed to God's.

## Apostasy: Denying Christ's Divinity

The danger doesn't end with moral compromise—it pierces to the heart of the gospel. In some circles, this drift has gone so far that the divinity of Jesus Christ is being minimized—or even denied.

That's not a minor error.

That's apostasy.

To reject Christ's divine nature is to cut off the foundation of salvation. Without the eternal Son of God, there is no Savior—only a hollow religion, stripped of power and hope.

These tragic compromises are the inevitable result of abandoning the authority of Scripture—beginning in Genesis.

Once that foundation cracks, even the identity of Christ becomes negotiable.

This is exactly what the apostles warned would happen: a slow erosion of truth that eventually leads to rejecting the Author Himself.

> *If the foundations are destroyed, what can the righteous do? — Psalm 11:3*

## Biblical Warnings About Apostasy

Scripture is clear: in the last days, many will fall away from the faith. This falling away is marked by a rejection of sound doctrine and an embrace of deception.

> *The Spirit expressly says that in latter times some will depart from the faith, giving heed to deceiving spirits and doctrines of demons. — 1 Timothy 4:1*

These warnings are not hypothetical,

They are happening *now.*

As churches compromise on foundational truths, many believers are drifting. What begins as cultural accommodation quickly becomes theological collapse.

Paul saw this coming:

> *For the time will come when people will not put up with sound doctrine. Instead, to suit their own desires, they will gather around them a great number of teachers to*

*say what their itching ears want to hear. — 2 Timothy 4:3*

This is the tragedy of our time:

A Church tempted to trade truth for relevance, and conviction for comfort. But truth was never meant to conform to culture.

*Truth is meant to confront—and transform—culture.*

When we stop contending for sound doctrine, we begin to blend in with the world—and lose our power to reach it.

## Jesus and the Narrow Gate

Jesus presents a sobering truth: there are only two paths in life—one that leads to destruction, and one that leads to life.

> *Enter through the narrow gate. For wide is the gate and broad is the road that leads to destruction, and many enter through it. But small is the gate and narrow the road that leads to life, and only a few find it. — Matthew 7:13*

This isn't a call to casual belief.

It's a warning—and a choice.

The broad road is easy, popular, and comfortable. It goes with the flow of culture, avoids hard truths, and demands nothing. But in the end, it leads to ruin.

The narrow road is different.

It's the path of truth, conviction, and surrender. It calls for repentance, obedience, and faith in God's Word—even when that Word is mocked or rejected.

Following this path often comes with a cost: rejection, ridicule, or even persecution. But it's the only road that leads to eternal life.

In a world rushing toward the broad road, Jesus calls His followers to walk the narrow road—to live faithfully, even when it's steep, unpopular, or lonely.

## The Narrow Gate Begins at Genesis

The narrow road doesn't begin in the Gospel of Matthew—it starts in Genesis. To walk the narrow road means trusting God's Word from the very first verse—not just when it's easy or socially acceptable. That includes believing that Genesis is literal history, not poetic metaphor or symbolic myth.

At the core, there are only two choices:

1. **Literal Genesis and a Good Creator**: Accepting Genesis as literal truth affirms God's goodness, justice, and the cause-and-effect nature of sin and redemption. It means believing creation was exactly as God described: intentional, good, and completed by His design.

2. **Theistic Evolution and a Contradictory Faith**: Blending Scripture with evolution leads to theological confusion. It

reshapes God's character, distorts the meaning of sin and the cross, and ultimately undermines the trustworthiness of Scripture—starting from page one.

Wrestling with these issues doesn't mean your faith is weak—it means you're seeking truth. And God invites us to seek Him—even in our doubts. His Word can stand up to your questions.

It is ***true, reliable, and worthy of your full trust.***

# AFTER EDEN
by Dan Lietha

Gap theory

161

Day-age theory

www.AnswersInGenesis.org

Theistic evolution

© 2003 AiG

Genesis 3:1

## In the beginning, the serpent in Eden was 'crafty' with God's Word. Since then, we've been crafty too.

Though this chapter focuses primarily on theistic evolution, the principle applies to other popular reinterpretations of Genesis— such as the *Gap Theory* or the *Day-Age Theory*. While these views differ in detail, they seem to share the same core flaw: they compromise the plain reading of Scripture to fit secular assumptions.

## Jesus and the Foundation of Creation

If we trust God's Word, we must also take seriously what Jesus Himself said about creation.

> *...at the beginning, the Creator made them male and female. — Matthew 19:4*

This wasn't symbolic. Jesus was affirming the literal creation of humanity—male and female—from the beginning. Not after millions of years of evolution, but from the foundation of the world.

And Jesus didn't speak as a distant commentator. He is the **Word made flesh** (John 1:1, 14) —the eternal Son of God and the very **Creator of all things**.

> *Through Him [Jesus] all things were made; without Him [Jesus] nothing was made that has been made. — John 1:3*

The apostle Paul affirms this again:

> *For by Him all things were created, in heaven and on earth... all things were created through Him and for Him. — Colossians 1:16*

So when Jesus speaks about creation, it's not as an observer, but as the *Author of creation itself.* He was there in the beginning, and He speaks with divine authority—the very wisdom that spun galaxies into being and breathed life into dust.

To question Jesus' understanding of creation is to question the One who created it. In a world where truth is often twisted, God's Word stands firm. And Jesus—the *Creator and Redeemer*—has authority over both the beginning of life and the path to eternal life.

To truly know Christ, we must believe what He believed— even about the beginning.

*There is no true gospel without a true Genesis.*

## Evolution vs. Human Dignity

If Jesus is our Creator, then every human being is *fearfully and wonderfully made*—not a cosmic accident, but a divine masterpiece crafted with purpose. The bible doesn't just tell us **how** we were made. It tells us **why** *we matter*.

We are made in the image of God.

> *So God created mankind in His own image… male and female He created them. — Genesis 1:27*

This truth is the foundation of human dignity, equality, and value. But evolution tells a very different story.

According to evolution, humanity is the result of a random, competitive process—governed by chance mutations and the survival of the fittest. In this worldview, some lives are naturally "more fit" or "more valuable" than others. History has shown the dangerous consequences of that thinking.

But the Bible insists that all human beings—regardless of race, age, ability, or background—have equal value because they bear God's image.

If we lose the doctrine of creation, we don't just lose theology—*we lose our humanity.*

Evolution strips away intrinsic worth, reducing people to biological accidents. Scripture, however, restores that worth, reminding us that each life is designed, known, and loved by a Creator.

Sadly, when a society forgets this, it opens the door to darkness:

- Eugenics
- Racism
- Abortion
- Genocide

Many of these horrors were supported—at least in part—by evolutionary thinking.

Only a worldview grounded in the image of God can uphold true justice, compassion, and human rights.

## Have You Ever Wondered Why We Die?

Beyond undermining human worth, evolution offers only a cold, mechanistic answer to one of life's most haunting questions: *Why do we die?*

From an evolutionary perspective, death is simply the result of natural processes—the victory of chemistry over biology. It's an impersonal outcome with no meaning or moral weight.

Even within that worldview, the question remains: If evolution is true, why hasn't nature evolved a way to eliminate death? Why does death remain an inescapable part of life?

The Bible gives a very different answer. It tells us that death is *not natural*—it's the tragic result of sin.

> *Therefore, just as sin entered the world through one man, and death through sin, and in this way death came to all people, because all sinned. — Romans 5:12*

This truth is reinforced just two verses later:

> *Nevertheless, death reigned from the time of Adam... — Romans 5:14*

Death entered the world through Adam's rebellion—when humanity first chose to disobey God and go its own way.

This biblical understanding of death is not only spiritually profound—it is essential to the gospel.

Because if sin brought death, only Christ's righteousness can bring life.

## The Story Doesn't End with Adam

The good news of the gospel is that Jesus came to undo what Adam did.

> *For as in Adam all die, so in Christ all will be made alive. — 1 Corinthians 15:22*

Through His sinless life, sacrificial death, and victorious resurrection, Jesus defeated death—offering eternal life to all who believe.

Evolution leaves us with despair.

The gospel leaves us with hope.

One sees death as natural and meaningless.

The other sees death as the enemy—and points to the One who conquered it.

*And because He lives, death doesn't get the final word.*

## Two Roads: One Leads to Life

The Bible's message is often hard to accept because it calls for humility, self-denial, and repentance—virtues that clash with our natural desires.

By contrast, the worldview of evolution offers a more comfortable message:

No absolute truth. Few moral boundaries. No higher power to answer to.

In that light, many adopt the philosophy:

*"If it feels good, do it."*

It promises freedom—but neglects the soul and the consequences that follow.

But Jesus paints a very different picture of life's path:

> *Whoever wants to be my disciple must deny themselves and take up their cross and follow me. — Matthew 16:24*

Following Jesus means embracing truth—even when it's uncomfortable. While the world often suppresses truth, Jesus calls us into light and freedom from sin's bondage.

Yet He also warns:

> *The wrath of God is being revealed from heaven against all the godlessness and wickedness of people,*

*who suppress the truth by their wickedness. — Romans
1:18*

This is why the message of the cross is urgent.

We are not spiritually neutral.

Human nature tends to suppress truth, not seek it.

That's why Jesus speaks of two roads—not three:

- One is broad and easy, **but leads to destruction.**

- The other is narrow and costly, **but leads to life.**

Only God's truth, received in humility, can lead us from
deception to salvation.

## The Urgency of Reconciliation with God

Imagine someone on the street being handed a vial of life-
saving medicine. But because they *feel fine*, they toss it aside. No
big deal, right?

Now imagine that same person had just been told by a doctor
that they had a rare, fatal disease. Suddenly, that medicine
becomes everything.[1]

That's salvation. Unless we recognize that we are spiritually
lost, we won't see the need for the cure—Jesus, the Great
Physician.

God won't force anyone into heaven; He respects our free will.
Yet many reject His invitation, choosing eternal separation instead.

Hell isn't God's desire. It's the consequence of refusing the only remedy.

God offers every person a choice: Follow Him or turn away.

That choice carries eternal consequences.

> *Here I am! I stand at the door and knock. If anyone hears my voice and opens the door, I will come in and eat with that person, and they with me. — Revelation 3:20*

He patiently knocks, waiting, loving, and longing for us to respond.

> *The Lord is not slow in keeping his promise, as some understand slowness. Instead, he is patient with you, not wanting anyone to perish, but everyone to come to repentance. — 2 Peter 3:9*

God's patience is a gift.

But it's also a call to response—now.

None of us is promised tomorrow. Once we take our last breath, the chance to be reconciled is gone.

*The time to seek Him is now—while the door is still open.*

## The Hidden Stumbling Block: Pride

Pride convinces us that we don't need God. It blinds us to our sin, and to our desperate need for grace.

It's easy to believe we have life under control, that we're good enough, or that our efforts can earn us favor with God. But pride distorts our vision, deceiving us about who we are and who God is. Yet how often do we really see ourselves as we are—apart from God?

Like a farmer must soften soil before planting, our hearts must be humbled to receive the gospel. We can't truly seek a Savior until we first admit we need one. Without this, salvation may seem unnecessary—or offensive. And for many, that's exactly the case.

From the start, pride caused humanity's fall. Adam and Eve weren't just tempted by fruit—they were tempted to become like God, knowing good and evil (Genesis 3:5). Their choice wasn't mere disobedience—it was defiant pride, believing they could define right and wrong apart from God.

That single act brought sin and death into the world—and that same prideful impulse still lives in every human heart.

Even today, pride keeps many from turning to the only One who can save.

## Pride's Insult: Why We Cannot Add to Christ's Work

Living as if God doesn't exist is the ultimate expression of pride. It says: *I know better*. It elevates human understanding above divine truth.

Pride tempts us to reinterpret Scripture through the lens of popular theories—like evolution—falsely exalting human wisdom.

Throughout Scripture, God makes one thing clear:

> *God opposes the proud but shows favor to the humble.*
> *— James 4:6*
>
> *Pride goes before destruction, a haughty spirit before a fall. — Proverbs 16:18*
>
> *For those who exalt themselves will be humbled, and those who humble themselves will be exalted. — Matthew 23:12*

Virtually every religion in the world teaches salvation or enlightenment through human effort—rituals, moral deeds, or performance. But biblical Christianity is radically different.

Scripture is clear: salvation is not "Christ plus works." It is a pure, unmerited gift—received by **grace alone,** through **faith alone,** in **Christ alone.**

God's plan of redemption was perfect from the start.

Christ executed it completely—and flawlessly.

To think we can add to His finished work is not only wrong—it's an insult to the sufficiency of His sacrifice. It is Christ alone—not Christ *plus* our baptism, sacraments, speaking in tongues, or spiritual performance. His finished work leaves no room for human merit:

*If it is by grace, then it is no longer by works; otherwise grace would no longer be grace. — Romans 11:6*

Pride may feel safe, but it leads to ruin.

Surrender may feel costly—but it leads to life.

*What good will it be for someone to gain the whole world, yet forfeit their soul? — Matthew 16:26*

Your soul is worth infinitely more than anything this world offers. No fleeting pleasure, title, or possessions can compare to the eternal joy and peace found in knowing Jesus.

## Religion vs. Relationship

The Bible warns us not to settle for lifeless religion—a form of godliness that looks holy on the outside but lacks true heart change.

*These people come near to me with their mouth and honor me with their lips, but their hearts are far from me. — Isaiah 29:13*

Outward words and rituals are not enough. What God truly desires is your heart.

True repentance goes beyond saying "sorry." It results in a transformed life—a change that becomes visible in how we worship and live. This transformation is not something we manufacture; it's the fruit of grace working within us. As Jesus said:

> *Yet a time is coming and has now come when the true worshipers will worship the Father in the Spirit and in truth. — John 4:23*

God is not after empty ritual—He longs for a real, honest relationship grounded in truth.

Sadly, for many, "religion" brings memories of pain rather than peace.

Some carry wounds from hypocritical leaders, abusive institutions, or legalistic systems that pushed them away from God instead of drawing them toward Him.

But those experiences do not reflect God's heart. In fact, His heart aches over every distortion of His love and truth.

Jesus reserved His strongest rebukes for the religious elite of His day—those who used their authority to burden people with rules rather than free them with grace:

> *Woe to you, teachers of the law and Pharisees, you hypocrites! You shut the door of the kingdom of heaven in people's faces. — Matthew 23:13*

The failures of religious people—even those in leadership—should never be mistaken for the character of God.

He is a loving, faithful Father who desires a genuine relationship with you—not a checklist of religious performance.

God didn't send His Son to start a religion.

*He sent Him to restore a relationship.*

## A Father's Heart

Imagine a father—not perfect, but deeply loving.

A father who would gladly give his life for his children.

He listens to them, delights in their laughter, and longs to spend time with them—not because he needs anything from them, but simply because he loves them.

He provides for them, builds a home for them, and pours out his heart every single day.

But one by one, his children turn away.

Some ignore him. Others mock him. A few even claim he never existed.

All he ever wanted was a relationship—but they wanted independence.

So he stands at the door of their hearts—knocking, waiting, loving—yet they refuse to let him in.

Despite everything, he doesn't walk away.

He waits.

***If you saw this father, wouldn't your heart break for him?***

And yet, this is just a glimpse of the love of God—and how deeply His heart aches when we reject Him. God didn't create us as robots, programmed to obey. He made us with the capacity for relationship—real love that must be freely chosen. But tragically, many choose distance over closeness, rebellion over restoration.

Still… even in rejection… the Father waits.

Still knocking.

Still loving.

Still hoping for one open door.

## The Simplicity of Salvation

One of the most beautiful—and astonishing—truths about Christianity is how simple it is to begin a relationship with God. It's not easy—but it is simple.

It's not about jumping through religious hoops, or trying to earn God's approval through good behavior.

God doesn't ask us to clean ourselves up before coming to Him—**He simply invites us to come as we are.**

Consider the thief on the cross beside Jesus.

He couldn't turn his life around, attend synagogue, or undo the wrongs he had done.

His hands were nailed in place—he had absolutely nothing to offer.

All he had was *repentance* and *faith.*

At first, both criminals mocked Jesus (Matthew 27:44), but something changed. As the hours passed, one thief had a moment of clarity. He realized his true condition—and more importantly, he realized who Jesus was.

With his dying breath, he acknowledged his guilt:

> *We are punished justly, for we are getting what our*
> *deeds deserve. But this man has done nothing wrong. —*
> *Luke 23:41*

That was **confession.**

That was **repentance.**

And then came faith:

> *Jesus, remember me when you come into your kingdom.*
> *— Luke 23:42*

He didn't have the right words. He didn't pray a polished prayer—just an honest one.

He didn't even ask for paradise—just to be remembered.

But in that broken, honest plea was the very essence of saving faith: **a repentant heart, fully trusting in Jesus.**

And how did Jesus respond?

> *Truly I tell you, today you will be with me in paradise.*
> *— Luke 23:43*

Not tomorrow.

Not after purgatory.

Not after proving himself.

**Today.**

That's grace.

**Undeserved. Unearned. Unshakable.**

That moment on the cross reveals the heart of the gospel:

It's not about what you've done—it's about whom you trust.

The doorway to salvation is open to *anyone* who repents and believes.

The thief had nothing but Jesus—*and it was more than enough.*

## Salvation is a Gift, Not a Reward

The thief didn't earn paradise. And neither can we.

The good news of Christianity is this:

Salvation isn't a reward for the righteous—it's a gift for the guilty.

God doesn't grade on a curve. He's not waiting for us to rack up spiritual points or outweigh our sins with good deeds. He knows we can't save ourselves. *That's why Jesus came.*

It's simple—but not shallow.

Grace cost Jesus everything, so that it could cost us nothing.

*For it is by grace you have been saved, through faith—*
*and this is not from yourselves, it is the gift of God—not*
*by works, so that no one can boast. — Ephesians 2:8–9*

The thief on the cross is living proof that it's never too late, you're never too far gone, and God's grace is always greater than your past.

*That gift is waiting—for you, too.*

## The Reality of Sin

The Bible is clear: **we are all sinners.**

*For all have sinned and fall short of the glory of God.*
*— Romans 3:23*

It's easy to feel "pretty good," when we compare our lives to others. But when we measure our lives against **God's holiness**, we all fall short. He sees every selfish thought, every prideful motive, every hidden sin.

When we honestly examine our lives through the lens of God's perfect law—His Ten Commandments—we come face to face with the truth:

- Have you ever told a lie? **That makes you a liar.**
- Have you ever stolen anything, no matter how small? **That makes you a thief.**

- Have you ever looked at someone with lust? **Jesus says that's adultery in your heart.**

- Have you ever harbored hatred or slandered someone? **God equates that with murder.**[2]

God's law wasn't given to prove how good we are—it exposes our condition so we'll seek a cure. It's a mirror, not a measuring stick. It's meant to point us to the cross.

Because God is holy, He cannot overlook sin. Because He is just, He cannot leave it unpunished.

Justice demanded payment.

Love stepped in—and paid it.

Jesus is God in the flesh—fully God and fully man. He willingly took our punishment on the cross. He was flogged, crucified, and killed. But on the third day, **He rose again— offering eternal life to all who believe.**

His work is finished. There is no other way to be saved— except through Him:

> *I am the way and the truth and the life. No one comes to the Father except through me. — John 14:6.*

Jesus' invitation is breathtakingly wide—but unmistakably narrow.

**It's wide**—because it's open to everyone, no matter your past.

**It's narrow**—because there's only one way: *through Him.*

> *Salvation is found in no one else, for there is no other name under heaven given to mankind by which we must be saved. — Acts 4:12*

## Responding to God's Invitation

God has made the way back to Him clear, but He won't force you. He lovingly invites you to come just as you are—*no cleanup required.*

You don't need religious rituals.

You need a relationship.

> *Today is the day of salvation. — 2 Corinthians 6:2*

If you're ready to begin a personal relationship with Jesus, here's how to start:

1. **Pray and Invite Jesus into Your Life**: Prayer is simply talking with God. If you're ready to surrender your will and heart to Him, you can pray something like this:

*Jesus, I believe that You are the Son of God and that You died for my sins. I confess that I have fallen short and need Your forgiveness. I ask You to come into my life, be my Savior and Lord, and help me to live for You from this day forward. Amen.*

It's not the words that save you—it's the faith behind them. If your heart is sincere, He hears you.

2. **Start Reading the Bible**: Begin with the New Testament, especially the *Gospel of John*. God's Word is living and powerful—it will change you from the inside out.

3. **Find a Bible-Based Church**: Look for a church that teaches the truth of God's Word. Surround yourself with others who are committed to following Him. You were never meant to walk this journey alone.

4. **Be Baptized**: Baptism doesn't save you—but it is a powerful, public declaration of your faith, symbolizing your new life in Christ.

5. **Stay Committed and Keep Growing**: Following Jesus is not a one-time decision, but a lifelong commitment. Continue praying, reading the Bible, and letting the Holy Spirit to shape you day by day.

## Final Encouragement

If you've placed your trust in Christ—you've made the most important decision of your life. Heaven is rejoicing—and so are we.

> *There is rejoicing in the presence of the angels of God over one sinner who repents. — Luke 15:10*

## The Gift of Eternal Life

> *For God so loved the world that He gave His one and only Son, that whoever believes in Him shall not perish but have eternal life.* — John 3:16

God didn't have to die for you—but He chose to, because He loves you that much.

Salvation is a gift. And like any gift, it must be received. When you turn from sin and place your trust in Jesus, your life is transformed. You are:

- Forgiven of every sin
- Spiritually reborn
- Adopted into God's family
- Brought into an eternal relationship with Him.

Don't wait. Come to Jesus today. His nail-pierced hands are open wide—ready to receive you.

He's not asking you to be perfect—just willing.

## Will You Open the Door?

Jesus is knocking on the door of your heart. *Will you open it?*

Will you trust Him—not just with your questions, but with your soul?

- Eternity is too important to ignore.

- Truth is too urgent to delay.

- Jesus is too loving to leave you where you are.

Today can be the day everything changes.

- Out of confusion... into clarity.

- Out of guilt... into grace.

- Out of death... into life.

*Then you will know the truth, and the truth will set you free. — John 8:32*

You don't need all the answers. You just need to trust the One who **is** the Answer. There is no reason to wait.

Some people reject the gospel on ANY grounds!

Too often, people reject the only lifeline that can save them—because it doesn't look the way they expected.

But Jesus is still reaching out. Will you take His hand?

**Will you surrender your heart to the One who:**

- Made you

- Died for you, and

- Longs to walk with you—now and forever?

Right now, Jesus is ready to change your *eternity*.

Your story doesn't have to end in separation.

It can begin in salvation.

**Right here. Right now. With Jesus.**

# Illustration Credits

Figures 1, 2, 3, 4, 5, 7, 8, 9, 10, 11, 12, 13, 14, 15, 17, 18, 19, 23, 24, 25, 26, 27, and 30:

Reproduced with special permission, courtesy of Richard Gunther — www.mightymag.org

Figure 6:

Image titled *"The Stones Will Cry Out,"* creator unknown. Widely circulated online. Used under fair use for educational and commentary purposes.

Figure 16:

Access Research Network. Copyright © 1998; All rights reserved. International copyright secured. File Date: 6.10.98; www.arn.org

Figure 20:

Reproduced with the permission of Natural Resources Canada 2012, courtesy of the Geological Survey of Canada (Photo 180345 by R.A. Price)

Figure 21: Institute for Creation Research

Figure 22: Rygel M.C

Figure 28 & 29: Answers in Genesis

# Notes

**Introduction: A personal Search for Truth**

[1] George Wald, "The Origin of Life," *Scientific American*, 191, no. 2 (August 1954): 48.

[2] Ben Stein, *Expelled: No Intelligence Allowed*, directed by Nathan Frankowski (Los Angeles: Premise Media, 2008).

[3] Rebecca Keller, "If We, as Scientists, Are Not Allowed to Question...," *Dissent from Darwin* (Discovery Institute), accessed May 14, 2025, www.dissentfromdarwin.org.

[4] *Dissent from Darwin*, accessed May 14, 2025, www.dissentfromdarwin.org.

[5] Dr. Stanley Salthe, "Darwinian Evolutionary Theory Was My Field of Specialization," *Dissent from Darwin* (Discovery Institute), accessed May 14, 2025, www.dissentfromdarwin.org.

[6] Douglas Axe, *A Scientific Dissent from Darwinism*, Discovery Institute, accessed May 14, 2025, www.dissentfromdarwin.org.

[7] Ken Ham, *The Lie: Evolution,* (Green Forest, AR: Master Books, 2009), 9. Used with permission from the publisher.


**Micro vs. Macro Evolution: What's the Difference?**

[1] David A. DeWitt, "Greater Than 98% Chimp/Human DNA Similarity? Not Any More: A Common Evolutionary Argument Gets Reevaluated—By Evolutionists Themselves," *Creation

*Ministries International*, accessed June 5, 2025, https://creation.com/greater-than-98-chimp-human-dna-similarity.

[2] Dawkins, Richard. *The Greatest Show on Earth: The Evidence for Evolution*. New York: Free Press, 2009.

[3] Vance Ferrell, *Science vs. Evolution* (Altamont, TN: Evolution Facts, Inc., 2006), chap. 11, 384.

[4] Lane P. Lester, "Genetics: no friend of evolution," *Creation* 20, no. 2 (March 1998): 20–22.

[5] John C. Sanford, *Genetic Entropy and the Mystery of the Genome* (Lafayette, LA: Ivan R. Dee, 2005).

[6] Georgia Purdom, "Is Natural Selection the Same Thing as Evolution?" in *The New Answers Book 1*, ed. Ken Ham (Green Forest, AR: Master Books, 2006), 271–282.

[7] Babu G. Ranganathan, "*Babu G. Ranganathan's Blog*," accessed December 2011, http://bgrnathan.blogspot.com.

[8] Lee Strobel, *The Case for a Creator* (Grand Rapids, MI: Zondervan, 2004), back cover.

**Fossilized Silence: What the Fossil Record Isn't Saying**

[1] Charles Darwin, *The Origin of Species* (1st ed.; New York: Avenel Books, 1979), 292.

[2] Luther D. Sunderland, *Darwin's Enigma: Ebbing The Tide of Naturalism*, 2nd printing 2002; Master Books, Inc., Green Forest, AR; ©1988, Ch 1, 11. Used with permission from publisher.

[3] Stephen C. Meyer, *Darwin's Doubt: The Explosive Origin of Animal Life and the Case for Intelligent Design* (New York: HarperOne, 2013).

[4] Jonathan Sarfati, *Refuting Evolution* (Green Forest, AR: Master Books, 2009), 32.

[5] Alan Feduccia, *The Origin and Evolution of Birds* (New Haven: Yale University Press, 1996), 123.

[6] Luther D. Sunderland, *Darwin's Enigma: Ebbing the Tide of Naturalism* (Green Forest, AR: Master Books, 1998), 101–2.

[7] David B. Kitts, "Paleontology and Evolutionary Theory," *Evolution* 28, (1974): 467

**The Cambrian Explosion: Evolution's Big Bang**

[1] Jeffrey S. Levinton, "The Big Bang of Animal Evolution," Scientific American, 267 (November 1992): 84.

[2] Richard Dawkins, *The Blind Watchmaker* (London: W.W. Norton & Company, 1987), chap. 9, 229.

[3] Charles Darwin, *The Origin of Species by Means of Natural Selection, or the Preservation of Favoured Races in the Struggle for Life*, 6th ed. (London: John Murray, 1872), 286.

[4] David M. Raup, *Extinction: Bad Genes or Bad Luck?* (New York: W.W. Norton, 1991), 3.

## Deception in the Details

[1] *Webster's New World Dictionary*, 3rd college ed. (New York: Simon & Schuster, 1988), s.v. "Piltdown Man.

[2] Chuck Colson, "The Archaeoraptor Fraud: This Bird Will Never Fly," *BreakPoint*, January 28, 2000, https://www.colsoncenter.org.

[3] Hillary Mayell, "Dino Hoax Was Mainly Made of Ancient Bird, Study Says," *National Geographic News,* November 20, 2012, http://news.nationalgeographic.com/news/2002/11/1120_021120_r aptor.html.

[4] Duane T. Gish, *Evolution: The Fossils Still Say No!* (El Cajon, CA: Institute for Creation Research, 1995), 126.

[5] Richard Leakey, quoted in *Creation Magazine* 12(3):32, June 1990, Creation Ministries International, https://creation.com/creation-magazine-12-3.

[6] Charles E. Oxnard, Fossils, Teeth and Sex: New Perspectives on Human Evolution (Seattle: University of Washington Press, 1987), p. 227.

[7] Trinkaus, Erik. "Modern Humans, Not Neandertals, May Be Evolution's 'Odd Man Out'." *ScienceDaily*, Sept. 8, 2006, https://www.sciencedaily.com/releases/2006/09/060908093606.htm.

## Dinosaurs: A Fresh Look

[1] "Komodo Dragon," *Encyclopædia Britannica*, accessed May 15, 2025, https://www.britannica.com/video/why-is-the-komodo-dragon-called-a-dragon/-321688.

[2] Dr. Leland Niermann, "Dinosaurs and Dragons," *Creation Ex Nihilo Technical Journal 8, no. 1 (1994): 85-104.*

[3] Peter Dickinson, *The Flight of Dragons* (New York: Harper & Row, 1979), 127.

[4] "Bushmen's Paintings Baffling to Scientists," *Los Angeles Herald Examiner*, 7 January 1970, quoted in Henry M. Morris, *That Their Words May Be Used Against Them* (San Diego: Institute for Creation Research, 1997), 252.

[5] Snelling, Andrew A. "Grand Canyon: Monument to Ancient Earth, Deceptions." *Answers Research Journal, Accessed June 30, 2025*.

[6] Dillow, J. C. *The Waters Above*. Chicago: Moody Press, 1981, 405–406. Quoted in "Will the Real Dr Snelling Please Stand Up?," *NoAnswersInGenesis*, accessed June 30, 2025.

[7] Brian Thomas and Vance Nelson, "Radiocarbon in Yet Another Dinosaur Fossil," *Creation Science Update*, Institute for Creation Research, July 18, 2019.

[8] Schweitzer, Mary H., Jennifer L. Wittmeyer, John R. Horner, and Jan K. Toporski. "Soft-Tissue Vessels and Cellular Preservation in *Tyrannosaurus rex*." *Science* 307, no. 5717 (2005): 1952–55. https://doi.org/10.1126/science.1104669.

[9] Alida M. Bailleul et al., "Evidence of proteins, chromosomes and chemical markers of DNA in exceptionally preserved dinosaur cartilage," *National Science Review* 7, no. 1 (2020): 110–21, https://doi.org/10.1093/nsr/nwaa007.

[10] Roach, John. "Dinosaur Mummy Found; Has Intact Skin, Tissue." *National Geographic*, 3 Dec. 2007, https://www.nationalgeographic.com/science/article/north-dakota-dinosaur-mummy.

[11] Yong, Ed. "Dinosaur proteins, cells and blood vessels recovered from Brachylophosaurus." *National Geographic*, 30 April 2009, https://www.nationalgeographic.com/science/article/dinosaur-proteins-cells-and-blood-vessels-recovered-from-bracyhlophosaurus.

[12] Mary H. Schweitzer et al., "A Role for Iron and Oxygen Chemistry in Preserving Soft Tissues, Cells and Molecules from Deep Time," *Proceedings of the Royal Society B: Biological Sciences* 281, no. 1775 (January 22, 2014): 20132741, https://doi.org/10.1098/rspb.2013.2741.

[13] Thomas G. Kaye, Gary Gaugler, and Zbigniew Sawlowicz, "Dinosaurian Soft Tissues Interpreted as Bacterial Biofilms," *PLOS One* 3, no. 7 (July 30, 2008): e2808, https://doi.org/10.1371/journal.pone.0002808.

## The Mousetrap Challenge: Do All the Parts Matter?

[1] Michael Behe, *Darwin's Black Box: The Biochemical Challenge to Evolution* (New York: Free Press, 1996), 45.

[2] *Unlocking the Mystery of Life*. (2002). *Documentary film*. Illustra Media.

[3] William A. Dembski, "Still Spinning Just Fine: A Response to Ken Miller," *DesignInference.com*, February 17, 2003, http://www.designinference.com/documents/2003.02.Still_Spinning.htm.

[4] Charles Darwin, *On the Origin of Species by Means of Natural Selection, or the Preservation of Favoured Races in the Struggle for Life* (London: John Murray, 1859), chap. 6, 189.

[5] *Expelled: No Intelligence Allowed*, directed by Nathan Frankowski (Premise Media, 2008), DVD.

## Do Evolution's Mechanisms Deliver?

[1] Henry Morris and Gary Parker, *What is Creation Science?* (printed 1987), 103—4. Used with permission from publisher – Master Books, Green Forest, AR; ©1982.

[2] Vance Ferrell, *Science vs. Evolution* (Altamont, TN: Evolution Facts, Inc., 2006), 390, chap. 11, third printing.

[3] Richard Dawkins, *The Greatest Show on Earth: The Evidence for Evolution (*New York: Free Press, 2009), 31.

[4] A. H. Linton, "Scant Search for the Maker," *Times Higher Education*, April 20, 2010, accessed November 9, 2010, timeshighereducation.co.uk.

[5] Spetner, Lee. *Not by Chance: Shattering the Modern Theory of Evolution*. Brooklyn, NY: Judaica Press, 1997, 138.

## Can Life Arise from Non-Life?

[1] James Perloff, *Tornado in a Junkyard: The Relentless Myth of Darwinism* (Burlington, MA: Refuge Books, 1999), 55–57.

[2] Sir Fred Hoyle, as quoted in "Hoyle on Evolution," Nature 294, no.5837 (November 12, 1981): 105.

[3] Francis Crick, *Life Itself: Its Origin and Nature* (New York: Simon & Schuster, 1981), 88.

[4] *Origins by Design* (Hagerstown, MD: Review and Herald Publishing Association, 1983), 376

[5] "The Big Bang in Astronomy," *New Scientist*, 92, no. 1280 (1981): 527.

[6] A. E Wilder-Smith, *The Natural Sciences Know Nothing of Evolution*, 4. Used with permission from the publisher – Master Books, Green Forest, AR; copyright©1981.

[7] Ibid., 9-89.

[8] Michael Denton, *Evolution: A Theory in Crisis* (Bethesda, Maryland: Adler and Adler Publishers, 1986), 264.

[9] James Perloff, *Tornado in a Junkyard: The Relentless Myth of Darwinism* (Arlington, MA: Refuge Books, 1999, Chap. 7, 70.

[10] Ibid., 70.

[11] Bruce A. Malone, *Search for the Truth: Changing the World with Evidence for Creation* (2006), 21.

[12] Tom Wagner, "A Squashed Mosquito is Dead Forever," *Creation* magazine, Answers in Genisis, www.answersingenesis.org.

[13] Michael J Behe, *The Edge of Evolution,: The Search for the limits of Darwinism*, Simon & Schuster (Free Press), 2007, back cover.

[14] James Perloff, *Tornado in a Junkyard: The Relentless Myth of Darwinism* (Arlington, MA: Refuge Books, 1999, Chap. 7, 74.

[15] George Sim Johnston, "Did Darwin Get It Right?" *The Wall Street Journal*, October 15, 1999.

[16] Walt Brown, *In the Beginning: Compelling Evidence for Creation and the Flood*, 8th ed. (Phoenix, AZ: Center for Scientific Creation, 2008), 3. Used with special permission. All rights reserved.

[17] From the book *In the beginning was information: A Scientist Explains the Incredible Design in Nature* by Werner Gitt, First printing, 2006; 11. Used with permission from the publisher - Master Books Inc., Green Forest, AR; copyright 2005.

[18] Ibid., 106.

[19] J.C. Sanford, *Genetic Entropy & The Mystery of the Genome*, 2nd ed. (Elim Publishing, 2005), 4.

**Five Scientific Laws Evolution Can't Escape**

[1] Halliday, David, Robert Resnick, and Jearl Walker. *Fundamentals of Physics*. 10th ed. Hoboken, NJ: Wiley, 2013, 182–184.

[2] Ross, Sheldon. *Introduction to Probability and Statistics for Engineers and Scientists*. 5th ed. New York: Elsevier, 2014, 137–140.

[3] Nagel, Ernest. *The Structure of Science: Problems in the Logic of Scientific Explanation*. New York: Harcourt, Brace & World, 1961, 22–24.

## *Can We Trust the Bible?*

[1] Josephus, *Antiquities of the Jews*, Book 18 and 20. English translation by William Whiston.

[2] John Warwick Montgomery, *History and Christianity* (Downers Grove, IL: InterVarsity Press, 1986), 25-26.

[3] Frank Turek and Norman Geisler, *I Don't Have Enough Faith to Be an Atheist* (Wheaton, IL: Crossway Books, 2004), 227-230.

[4] Lee Strobel, The Case for Christ (Grand Rapids, MI: Zondervan, 1997), 65-68.

[5] Norman L. Geisler and Ronald M. Brooks, *When Skeptics Ask: A Handbook on Christian Evidences* (Grand Rapids, MI: Baker Books, 1990), 164-165.

[6] Peter W. Stoner, *Science Speaks* (Moody Press, 1963), 100–107.

## The Case for a Global Flood

[1] Richard D. Draper, *Flood: In Search of the Origins of Noah's Flood* (Salt Lake City: Deseret Book, 2013).

[2] Snelling, Andrew A. *Earth's Catastrophic Past: Geology, Creation & the Flood*. 2 vols. Dallas, TX: Institute for Creation Research, 2009.

[3] John Morris, *The Global Flood: Ancient Legends and Modern Evidence* (Institute for Creation Research, 2015).

[4] Andrew George, *The Epic of Gilgamesh: The Babylonian Epic Poem and Other Texts in Akkadian and Sumerian* (London: Penguin Classics, 2003).

[5] Henry M. Morris, *The Genesis Flood* (Grand Rapids: Baker Book House, 1961).

[6] Brown, Walt. *In the Beginning: Compelling Evidence for Creation and the Flood*, 5th ed., chap. 7, "The Grand Canyon" (Phoenix, AZ: Center for Scientific Creation, 2008).

[7] Snelling, Andrew A. *Earth's Catastrophic Past: Geology, Creation & the Flood*. Dallas, TX: Institute for Creation Research, 2009.

[8] Hill, Andrew. "The Great Unconformity and the Cambrian Explosion: New Perspectives on Earth's History." *Journal of Creation* 29, no. 1 (2015): 73–83.

[9] Clarey, Timothy L. *Carved in Stone: Geological Evidence of the Worldwide Flood*. Dallas, TX: Institute for Creation Research, 2015.

[10] Powell, John Wesley. *Exploration of the Colorado River and Its Canyons*. Washington, D.C.: Government Printing Office, 1875.

[11] Isis Temple." *Wikipedia*. Last modified January 5, 2025. https://en.wikipedia.org/wiki/Isis_Temple.

[12] Havasupai Tribe. "Havasupai History and Legends." Havasupai Tribal Website, accessed May 2025. https://www.havasupai-nsn.gov/history.

[13] Nickel, Bryan. *Hydroplate Theory: Origins of the Grand Canyon*. YouTube video, 48:15. Posted August 2019. https://www.youtube.com/watch?v=Hqvroege-Hk.

[14] Institute for Creation Research. "Lessons from Mount St. Helens Eruption." Accessed May 28, 2025. https://www.icr.org/article/lessons-from-mount-st-helens-eruption/.

[15] Evolution of Thinking. "Mount St. Helens: A Living Research Laboratory." Accessed May 28, 2025. https://evolutionofthinking.org/earth/geology/mount-st-helens-a-living-research-laboratory/.

[16] Steven A. Austin, *Mount St. Helens and Catastrophism*, Acts & Facts (Institute for Creation Research), May 1986, https://www.icr.org/article/mount-st-helens-catastrophism/.

[17] Washington-Oregon, Caused by the May 18, 1980, Eruption of Mount St. Helens." Accessed May 28, 2025. https://pubs.usgs.gov/publication/cir850K.

[18] U.S. Senator Maria Cantwell, *Senate to Remember Mt. St. Helens Disaster*, press release, May 17, 2010, https://www.cantwell.senate.gov/news/press-releases/senate-to-remember-mt-st-helens-disaster.

[19] Steven A. Austin, "The Polystrate Trees and Coal Seams of Joggins Fossil Cliffs," *Institute for Creation Research*, accessed May 28, 2025, https://www.icr.org/article/polystrate-trees-coal-seams-joggins.

[20] Tas Walker, "Polystrate Fossils: Evidence for a Young Earth," *Creation Ministries International*, last modified October 25, 2006, https://creation.com/polystrate-fossils-evidence-for-a-young-earth.

[21] Wikipedia, s.v. "Polystrate fossil," last edited January 5, 2025, 10:26 (UTC), https://en.wikipedia.org/wiki/Polystrate_fossil.

[22] Institute for Creation Research. "Marine Invertebrates." *Institute for Creation Research*. Accessed May 28, 2025. https://www.icr.org/invertebrates/.

[23] Walter T. Brown, *In the Beginning: Compelling Evidence for Creation and the Flood* (Phoenix, AZ: Center for Scientific Creation, 2008).

[24] Baumgardner, John R. "14C Evidence for a Recent Global Flood and a Young Earth." In *Proceedings of the Fifth International Conference on Creationism*, edited by R. L. Ivey Jr., 127–142. Pittsburgh, PA: Creation Science Fellowship, 2003.

[25] Tim Clarey, *"Fossils Out of Order: Disrupting the Geologic Column,"* Answers in Genesis, accessed May 29, 2025, https://answersingenesis.org/fossils/fossil-record/fossils-out-of-order-disrupting-geologic-column/.

## A Universe Too Perfect to Be an Accident

[1] John D. Morris, *The Young Earth* (Green Forest, AR: Master Books, 2007), 58-61.

[2] Jonathan Sarfati, *The New Answers Book 1* (Green Forest, AR: Master Books, 2006), 120-125.

[3] Stephen C. Meyer, *Signature in the Cell: DNA and the Evidence for Intelligent Design* (New York: HarperOne, 2009), 72-85.

[4] Michael Behe, *Darwin's Black Box: The Biochemical Challenge to Evolution* (New York: Free Press, 1996), 120-130.

[5] Jonathan Sarfati, *Refuting Evolution 2* (Green Forest, AR: Master Books, 2010), 98-110.

[6] George Wald, *The Origin of Life*, Scientific American, 191:48. May 1954.

[7] Professor Louis Bounoure, Director of Research, National Center of Scientific Research, The Advocate, 8 March 1984.

[8] Dr. T. N. Tahmisian, "Scientists Who Go About Teaching That Evolution Is a Fact of Life Are Great Con-Men," *The Fresno Bee*, August 20, 1959, as quoted in N. J. Mitchell, *Evolution and the Emperor's New Clothes* (Roydon Publications, 1983), title page.

[9] Robert Jastrow, *God and the Astronomers* (New York: W. W. Norton, 1992), 107

## The Emperor's New Theory: Why Evolution Still Reigns

[1] Aldous Huxley, *Ends and Means* (London: Chatto & Windus, 1946), 270, 273.

[2] Richard Lewontin, "The Demon-Haunted World," *The New York Review of Books*, January 9, 1997, 28.

[3] *The Battle for the Beginning*, John MacArthur, copyright © 2007, p. 38, Thomas Nelson Inc. Nashville Tennessee. All rights reserved. Reprinted by permission.

## The Cultural Fallout of Evolution

[1] Charles Darwin, *The Autobiography of Charles Darwin 1809–1882*, ed. Nora Barlow (London: Collins, 1958), 86.

[2] Centers for Disease Control and Prevention (CDC), *Youth Suicide Rates: Trends and Analysis*, accessed April 25, 2023, https://www.cdc.gov/nchs/data/databriefs/db352.pdf

[3] Gun Clubs at School: Not so long ago, they were common — and safe." Free Republic. Accessed May 21, 2025. https://freerepublic.com/focus/news/2980293/posts.

[4] Arthur Keith, *Evolution and Ethics* (London: Williams & Norgate, 1947), as cited in Bert Thompson, *Biological Evolution* (Montgomery: Apologetics Press, 1990), 7.

[5] Adolf Hitler, *Mein Kampf*, trans. James Murphy, Chap. 11, "Race and People," http://gutenberg.net.au/ebooks02/0200601.txt.

[6] Yaroslavsky, E. *Landmarks in the Life of Stalin*. Moscow: Foreign Languages Publishing House, 1940.

[7] Tom DeRosa, *Evolution's Fatal Fruit*, foreword by D. James Kennedy (Coral Ridge Ministries, 2006), 8.

[8] James Kennedy (forward), Tom DeRosa, *Evolution's Fatal Fruit*, Coral Ridge Ministries, 2006, 14.

[9] DeRosa, Tom. *Evolution's Fatal Fruit: How Darwin's Tree of Life Brought Death to Millions*. Coral Ridge Ministries, 2006.

[10] Lauren Enriquez, "Even After Removing Margaret Sanger's Name, Planned Parenthood Is Still Influenced by Racist Eugenics," *The Heritage Foundation*, July 24, 2020, https://www.heritage.org/life/commentary/even-removing-margaret-sangers-name-planned-parenthood-still-influenced-racist.

[11] *Ibid*
[12] *Ibid*

## Evolution in Schools: Is There a Hidden Bias?

[1] *Torcaso v. Watkins*, 367 U.S. 488, at 495 n.11 (1961); *Smith v. Board of School Com'rs of Mobile County*, 655 F. Supp. 939 (S.D. Ala. 1987), *rev'd*, 827 F.2d 684 (11th Cir. 1987).

[2] Charles Francis Potter, *Humanism: A New Religion* (New York: Simon and Schuster, 1930), 128, quoted in David A. Noebel, J. F. Baldwin, and Kevin Bywater, *Clergy in the Classroom: The Religion of Secular Humanism* (Manitou Springs, CO: Summit Press, 1995), vi.

[3] John J. Dunphy, "A Religion for a New Age," *The Humanist*, January–February 1983, 26.

<sup>4</sup> *Ibid*

<sup>5</sup> *Reed v. Van Hoven*, 237 F. Supp. 48 (W.D. Mich. 1965).

<sup>6</sup> International Labour Organization (ILO), Walk Free, and International Organization for Migration (IOM), *Global Estimates of Modern Slavery: Forced Labour and Forced Marriage* (Geneva: ILO, 2022), 7.

<sup>7</sup> Engel v. Vitale, 370 U.S. 421 (1962); Abington School District v. Schempp, 374 U.S. 203 (1963).

<sup>8</sup> *Epperson v. Arkansas*, 393 U.S. 97 (1968).

<sup>9</sup> *Edwards v. Aguillard*, 482 U.S. 578 (1987).

<sup>10</sup> *Town of Greece v. Galloway*, 572 U.S. 565 (2014).

<sup>11</sup> *American Atheists, Inc. v. American Legion*, 895 F.3d 265 (4th Cir. 2018).

<sup>12</sup> *Wang v. Heffernan*, 912 F.3d 1169 (9th Cir. 2019).

<sup>13</sup> Thomas Jefferson, "Letter to the Danbury Baptists," January 1, 1802, in *The Writings of Thomas Jefferson*, vol. 16, ed. Albert Ellery Bergh (Washington, D.C.: Thomas Jefferson Memorial Association, 1907), 281-82.

## The Bible vs. Evolution: The Ultimate Collision

<sup>1</sup> Todd Friel, *Wretched: The 10 Best Things About God (and Why You'll Never Be the Same)* (Grand Rapids, MI: Zondervan, 2011).

<sup>2</sup> Comfort, Ray. *The Way of the Master*. Gainesville, FL: Bridge-Logos, 2004.

www.ingramcontent.com/pod-product-compliance
Lightning Source LLC
Chambersburg PA
CBHW071704120626
46550CB00001B/100